Fran Sérgio Lobato
Romes Antonio Borges
Valder Steffen Jr.

Bio-inspired Optimization Methods

Fran Sérgio Lobato
Romes Antonio Borges
Valder Steffen Jr.

Bio-inspired Optimization Methods

modeling, design, inverse problem and robust
optimization of a representative mechanical
system

LAP LAMBERT Academic Publishing

Imprint

Any brand names and product names mentioned in this book are subject to trademark, brand or patent protection and are trademarks or registered trademarks of their respective holders. The use of brand names, product names, common names, trade names, product descriptions etc. even without a particular marking in this work is in no way to be construed to mean that such names may be regarded as unrestricted in respect of trademark and brand protection legislation and could thus be used by anyone.

Cover image: www.ingimage.com

Publisher:
LAP LAMBERT Academic Publishing
is a trademark of
Dodo Books Indian Ocean Ltd. and OmniScriptum S.R.L publishing group

120 High Road, East Finchley, London, N2 9ED, United Kingdom
Str. Armeneasca 28/1, office 1, Chisinau MD-2012, Republic of Moldova, Europe
Managing Directors: Ieva Konstantinova, Victoria Ursu
info@omniscriptum.com

Printed at: see last page
ISBN: 978-3-659-81486-0

BIO-INSPIRED OPTIMIZATION METHODS - MODELING, DESIGN, INVERSE PROBLEM AND ROBUST OPTIMIZATION OF A REPRESENTATIVE MECHANICAL SYSTEM

Fran Sérgio Lobato[1], Romes Antonio Borges[2] and Valder Steffen Jr[3]

[1]Laboratory of Modeling, Simulation, Control and Optimization of Process, School of Chemical Engineering, Federal University of Uberlândia, Av. João Naves de Ávila, 2121, Uberlândia, MG, ZIP CODE 38400-902, Brazil, fslobato@feq.ufu.br

[2]Mathematics and Technology Institute - School of Industrial Mathematics, Federal University of Goiás, Av. Lamartine P. Avelar, 1120, Catalão, GO, ZIP CODE 75704-020, Brazil, romes@ufg.br

[3]Laboratory of Mechanics and Structures (LMEst), School of Mechanical Engineering, Federal University of Uberlândia, Av. João Naves de Ávila, 2121, Uberlândia, MG, ZIP CODE 38400-902, Brazil, vsteffen@mecanica.ufu.br

CONTENTS

PREFACE

Optimal robust design is an important topic for mechanical systems and structures. In the present contribution, the synthesis of a nonlinear dynamic vibration absorber is used as a representative example of mechanical system design. Dynamic vibration absorbers (DVA) are mechanical devices designed to attenuate the vibration level of different structures and machines used in several engineering applications, such as ships, power lines, aeronautic structures, civil engineering constructions subjected to seismic induced excitations, among others. Traditionally, different approaches have been proposed to DVA design, solve inverse problems and to obtain robust optimal solutions for DVA applications. In this contribution, a mathematical modeling and sensibility analysis of a two degree-of-freedom nonlinear damped system constituted of a primary mass attached to the ground by a linear spring and the secondary mass attached to the primary system by a nonlinear spring (nDVA) is presented. In the design context, the nDVA optimal is obtained by considering the maximization of the attenuation bandwidth together with the minimization of the system response. In addition, the solution of an inverse problem and the robust optimization problem are also addressed. For all applications conveyed, the optimization strategy used is based on Bio-inspired Optimization Methods in association with two particular operators, i.e., the Pareto Dominance Criterion and the Crowding Distance Operator.

Keywords: *Robust Optimization, Nonlinear Dynamic Vibration Absorber, Inverse Problem, Bio-inspired Optimization Methods.*

CHAPTER 1:
INTRODUCTION

1.1 MOTIVATION

The modeling, simulation, design, control and optimization of real-world engineering systems represent a difficult task due to the complexity of the phenomena involved. All these steps require high computational effort to find the solution of realistic problems. In the context of mechanical engineering design, engineers have to deal with nonlinear systems in which the dynamic response depends on a number of physical parameters. An example of such a system is the so-called dynamic vibration absorber (DVA). DVAs are mechanical appendages comprising inertia, stiffness and damping elements, which, once connected to a given structure or machine (primary structure), are used to reduce vibration (and noise) in various types of engineering systems, such as compressors, robots, ships, power lines, airplanes, and helicopters (Frahm, 1911; Hartog, 1934; Koronev and Reznikov, 1993; Steffen Jr and Rade, 2001). An important advantage of a DVA in comparison with other methods for vibration reduction is that it can be applied to structures that are already in operation and appear to have unsatisfactory dynamic responses. In addition, they are able to reduce the vibration level of a structure at a comparatively low cost (Bonsel, 2003), on a passive way.

In the last years, various efforts have been devoted to the development of mathematical models for characterizing the mechanical behavior of nonlinear dynamic vibration absorbers (nDVA) accounting for their typical dependence on design parameters that influence the nonlinear behavior of the system. Rice and McCraith (1987) and Shaw et al. (1989) suggested optimization strategies to be applied to the design of nDVA by applying an asymmetric nonlinear Duffing-type element incorporated in the suspension for narrow-band absorption applications. Espíndola and Bavastri (1997) studied a particular type of nDVA, the so-called viscoelastic neutralizer, and Espíndola et al. (2008, 2009) studied strategies for the design of viscoelastic vibration absorbers. Besides the well-known complexity of the modeling strategy involved in nonlinear dynamics, some general methodologies have been suggested and have shown to be particularly suitable to be used in combination with techniques of structural systems discretization. This aspect makes them very attractive for the modeling of nonlinear dynamic vibration absorbers, so that the stability and efficiency of nDVA into a frequency band of interest can be improved, leading to refined nDVA design (Nissen et. al., 1985; Pai and Schulz, 2000). Wong and Cheung (2008) proposed an optimum tuning condition including the frequency and damping ratios of the absorber based on the fixed-points theory. These authors proved analytically that the proposed absorber provide larger suppression of resonant vibration amplitudes of the primary system as excited by ground motion than the traditional absorber. Viana et al. (2008) determined the optimal tuning of two different types of dynamic vibration absorbers (DVA) by using ant colony optimization. The tuning of dynamic vibration absorbers is the procedure that sets the anti-resonance frequency to a given value by adjusting the parameters of the dynamic vibration absorber, considering the minimization of vibration amplitude of the primary structure. Borges et. al. (2010) determined the robust optimal design of an nDVA combining sensitivity analysis and multi-objective optimization. Yang et al. (2011) used a dynamic vibration absorber to suppress the vibration of a plate in a frequency band. The control mechanism is investigated with respect to different bandwidths, and the coupling is examined in terms of the resonator location and frequency bandwidth. Cheung et al. (2012) proposed the design of a hybrid vibration absorber for the minimization of the resonant vibration amplitude of a single degree-of-freedom vibrating structure by using the fixed-points theory. These authors analyzed the effects of the feedback gain, the tuning frequency ratio, the damping ratio and the mass ratio of the absorber on the vibration absorption of the primary structure. Borges et al. (2013) proposed the optimal design of a nonlinear optimization problem characterized by a two-degree-of-freedom nonlinear damped system. A primary mass attached to the ground by a linear spring and a secondary mass attached to the primary system by a nonlinear spring characterizes the system, for which three bio inspired optimization methods were tested at the design phase (Bees Colony Algorithm, Firefly Colony Algorithm and Fish Swarm Algorithm). Cheung et al. (2013) optimized a hybrid vibration absorber for suppressing resonant vibration of a single degree-of-freedom system under stationary random force excitation. The methodology proposed includes damping in the primary system and the effect of this damping on the optimization process is evaluated. It is found that optimum values of the system parameters do not exist if damping introduced in the primary system is relatively high. Febbo and Machado (2013) analyzed the dynamics of a novel nonlinear dynamic vibration absorber attached to a linear/nonlinear

3

primary system. A method of averaging was selected to obtain the frequency response curves, which proved to be appropriate to model the saturation phenomenon of the absorber.

A straightforward extension of the modeling techniques is the optimization of nDVA devices aiming at reducing the cost and/or maximizing their performance. In the quest for optimization, engineers are frequently faced with conflicting objectives, which naturally lead to the use of multi-criteria optimization procedures (Eschenauer et. al., 1990). In this context, an important topic to be addressed is the so-called sensitivity analysis, which enables to evaluate the degree of influence of variations of physical and/or geometrical parameters on the dynamic behavior of nDVAs. Sensitivity analysis represents an important step in various types of problems such as model updating, analysis of modified structures, optimal design, system identification, control, and stochastic reliability assessment (Lima et al., 2006). Several approaches have been developed for performing sensitivity analysis of dynamic responses, as reported in references (Haug et al., 2006; Lima et al., 2006). However, applications to the case of structural systems that contain nDVAs in which the analysis is combined with robust optimization strategies (Lee and Park, 1996ab) are not numerous, which motivates the study reported here.

Naturally, real-world optimization problems can be formulated by taking into account two or more (often conflicting) objectives. The approach to such problems (called multi-criteria or multi-objective optimization problems - MOOP) is different from the one that considers a single-objective optimization problem. The main difference is that multi-objective optimization problems normally have not a single solution, but, on the contrary, a set of solutions that are all equally satisfactory (Deb, 2001). In the literature, several methods for solving MOOP can be found. These follow a preference-based approach, in which a relative preference vector is used to scalarize multiple objectives. Since classical searching and optimization methods use a point-by-point approach, at which the solution is successively modified, the outcome of this classical optimization method is a single optimized solution (Deb, 2001). However, Evolutionary Algorithms (EA) can find multiple optimal solutions in one single simulation run due to their population-based search approach. Thus, EA are ideally suited for multi-objective optimization problems.

In this context, Bio-inspired Optimization Methods (BiOM) are configured as interesting approaches to solve both mono-objective and multi-objective problems. Fundamentally, these methodologies are based on strategies that seek to mimic the behavior observed in species found in the nature to update a population of candidates

4

to solve optimization problems (Yang, 2008). These systems have the capacity to notice and modify their environment in order to seek for diversity and convergence. In addition, this capacity makes possible the communication among the agents (individuals of population) that capture the changes in the environment as generated by local interactions (Yang, 2008). Among the most recent bio-inspired strategies, stand the Bees Colony Algorithm - BCA (Pham et. al., 2006), the Firefly Colony Algorithm - FCA (Yang, 2008), and the Fish Swarm Algorithm - FSA (Li et. al., 2002). The BCA is based on the behavior of bee colonies in their search of raw materials for honey production. According to Lucic and Teodorovic (2001), in each hive groups of bees (called scouts) are recruited to explore new areas searching for pollen and nectar. These bees, returning to the hive, share the acquired information so that new bees are indicated to explore the best regions visited in an amount proportional to the previously passed assessment. Thus, the most promising regions are best explored and eventually the least end up being discarded. Every iteration in this cycle repeats itself with new areas being visited by scouts. The FCA mimics the patterns of short and rhythmic flashes emitted by fireflies in order to attract other individuals to their vicinities. The corresponding optimization algorithm is formulated by assuming that all fireflies are unisex, so that one firefly will be attracted to all other fireflies. Attractiveness is proportional to their brightness, and for any two fireflies, the less bright will attract (and thus move to) the brighter one. However, the brightness can decrease as their distance increases and if there are no fireflies brighter than a given firefly it will move randomly. The brightness is associated with the objective function for optimization purposes (Yang, 2008). Finally, the FSA is a random search algorithm based on the behavior of fish swarm observed in nature. This behavior may be summarized as follows (Li et. al., 2002): random behavior - in general, fish looks at random for food and other companion; searching behavior - when the fish discovers a region with more food, it will go directly and quickly to that region; swarming behavior - when swimming, fish will swarm naturally in order to avoid danger; chasing behavior - when a fish in the swarm discovers food, the others will find the food dangling after it; and leaping behavior - when fish stagnates in a region, a leap is required to look for food in other regions.

An example of application of the BiOM is the solution of inverse problems. In the field of mechanical systems, this approach may be used for the estimation of boundary conditions, such as temperature and heat flux, determination of thermal-physical properties. This problem, also called as parameter identification problem, arise from the necessity of obtaining parameters of theoretical models in such a way that the models can be used to simulate the behavior of the system for different operating conditions. Basically, the estimation procedure consists in obtaining the model parameters through the minimization of the difference between calculated and experimental values considering a given optimization strategy. In this context, the BiOM in association with the inverse analysis can be used to determine parameters (proprieties) or boundary conditions in nDVA through of use of experimental data.

Traditionally during the solution of optimization problems it is commonly considered that mathematical models, variables and parameters are sufficiently reliable, i.e., there are no errors of modeling and estimation (Deb and Gupta, 2006). However, systems to be optimized are generally sensitive to small changes in the design variables leading to significant changes in the vector of objective functions. In this case, to minimize this effect in the solution of optimization problems, the concept of robust optimization should be considered. Robust Optimization is defined as an approach that produces a solution that is not sensitive to small changes in the design variables (Taguchi, 1984). In addition, this approach is used for modeling optimization problems under uncertainty, where the modeler aims at finding decisions that are optimal for the worst-case realization of the uncertainties within a given set of values (Taguchi, 1984).

The introduction of robustness in the mono and multi-objective context requires the consideration of new restrictions and/or new objectives (relations between the mean and the standard deviation of the objective functions vector) and probability distribution functions for the design variables and/or objectives. As mentioned by Ritto et al. (2008) and considered by Soize (2001, 2005), Cataldo et al. (2007, 2008, 2009) and Sampaio and Soize (2007), probability tools are used to model uncertainties, i.e., random variables are associated to the uncertain parameters or matrices and probability density functions are then constructed.

As an alternative to these classical formulations, Deb and Gupta (2006) extended the Mean Effective Concept (MEC), originally proposed for mono objective problems, for the multi-objective context. In this approach, no additional restriction is inserted into the original problem. Thus, the problem is rewritten as a mean vector of original objectives. These authors applied this methodology to constrained and unconstrained test problems having two and three objectives and showed interesting simulation results using an evolutionary multi-objective optimization algorithm and engineering design techniques. More recently, Souza et. al. (2015) applied this methodology to solve singular optimal control problems with different levels of complexity, where the original continuous control trajectory is approximated by linear functions on time intervals by using the Multi-objective Optimization Differential Evolution algorithm. As mentioned by the authors, it is important to observe that the main disadvantage of this approach is the increase of the number of objective function evaluations, necessary to evaluate the integral considered in the MEC formulation.

As observed in the literature, the development of new methodologies involving the nDVA represents a subject of permanent interest due to their technological relevance both in the academic and industrial domains. In this book, the modeling, design, solution of inverse problems, and robust optimization of a two degree-of-freedom nonlinear damped system, constituted of a primary mass attached to the ground by a linear spring and the secondary mass attached to the primary system by a nonlinear spring is presented.

In the book, the theoretical foundations related to various aspects regarding the strategy for modeling nDVAs, sensitivity analysis, inverse problem technique, and robust optimization procedures are first summarized, followed by the description of a numerical application that demonstrates the effectiveness of the methodology conveyed, when applied to obtain the robust optimal design of structural systems incorporating nDVAs. Chapter 2 addresses the mathematical modeling of Nonlinear Dynamic Vibration Absorber, the sensibility analysis and numerical simulation results for different system configurations. Chapter 3 presents the concept of BiOM (Bees Colony Algorithm, Firefly Colony Algorithm and Fish Swarm Algorithm), its extension to the multi-objective context and the design of an nDVA considering the design variables obtained from the sensibility analysis. Chapter 4 discusses the use of BiOM for the solution of an inverse problem (parameter identification) through the minimization of the difference between calculated and synthetic experimental values. The robust optimization is studied in the Chapter 5. Finally, the conclusions are outlined in this last chapter.

CHAPTER 2:
MATHEMATICAL MODELING NONLINEAR DYNAMIC VIBRATION ABSORBERS

2.1 INTRODUCTION

As mentioned above, nonlinear dynamic absorbers are considered as representative examples of mechanical systems. Mechanical systems subjected to different external forces and/or to a set of initial conditions, respond dynamically resulting a small amplitude oscillation around an equilibrium position. This movement is generally defined as "vibration", which is the result of an ongoing process of transformation involving the kinetic and potential energies of the system.

Damping is related to a mechanism of energy dissipation of mechanical systems. The most widely used models of damping that are incorporated to the systems involve the dry friction (Coulomb) and the viscous damping. Obviously, in the case of damped systems the mechanical energy of the system is not preserved; on the contrary, it continually decreases over time.

There are different ways of mitigating undesired vibrations in a given structure. The most traditional way relies on devices that are known as Dynamic Vibration Absorbers (DVA) (Steffen Jr and Rade, 2001). A dynamic vibration absorber usually consists of a mass that is coupled to the primary system, which is then tuned to absorb some of the vibrational energy of the system.

The use of discrete DVAs to the problem of vibration attenuation constitutes an important subject in modern Engineering. The application of DVAs to reduce noise and vibration in various types of engineering systems such as compressors, robots, ships, power lines, airplanes, helicopters, space structures, buildings and towers, etc., has been intensively investigated along several decades. Much of the knowledge available to date is compiled in the books by Frahm (1911) and Koronev and Reznikov (1993) and in some review papers such as those by Cunha Jr (1999), and Steffen Jr and Rade (2000, 2001).

In the last decades, a great deal of effort has been devoted to the development of mathematical models for characterizing the mechanical behavior of nonlinear dynamic vibration absorber (nDVA) accounting for its typical dependence on parameters that control the nonlinearities. Besides, the well-known complexity of the modeling strategy of the nonlinear dynamics, which constitutes a simple and straightforward means of representing the dynamic behavior of nDVAs, some strategies have been suggested and have been shown to be particularly suitable to be used in combination with structural systems discretization. This aspect makes them very attractive for the modeling of nonlinear dynamic vibration absorbers. Among those strategies, it should be mentioned the theoretical study proposed by Pai and Schulz (2000) in which strategies to improve the stability and efficiency of nDVAs into a frequency band of interest have been proposed, leading to a refined system containing nDVAs. In addition, Rice and McCraith (1987) suggested optimization strategies to be applied into the project of nDVAs by incorporating an asymmetric nonlinear Duffing-type element for narrow-band absorption applications.

A natural extension of modelling capability is the optimization of nDVAs aiming at reducing cost and/or maximizing their performance (Nissen et. al., 1985). In the quest for optimization, engineers are frequently faced with conflicting objectives, which naturally leads to the use of multi-criteria optimization procedures (Eschenauer et. al., 1990; Lima, 2007; Borges et. al., 2010).

In this context of analysis and design of structural systems incorporating nDVAs, an important topic to be addressed is the sensitivity analysis, which enables to evaluate the degree of influence of variations of physical and geometrical parameters that are responsible to the nonlinear behaviour of the system. Sensitivity analysis constitutes an important step for various types of problems such as model updating, analysis of modified structures, optimal design, system identification, control and stochastic reliability assessment (Haug et al., 1986).

Several approaches have been developed for performing sensitivity analysis of dynamic responses, as reported in the literature (Eschenauer et. al., 1990; Lee and Park, 1996ab). However, applications to the case of structural systems containing nDVAs components are not numerous, which motivates the study reported herein.

2.2 NONLINEAR DYNAMIC VIBRATION ABSORBER

Consider the vibratory system represented by the two degree-of-freedom model shown in Fig. 2.1 (Borges et. al., 2010). The interest is focused on frequency domain responses. In this case, to calculate the steady-state harmonic responses in the frequency domain the following relation for the external force $f(t) = f_0 \cos(\omega t)$ is assumed (where ω is the excitation frequency).

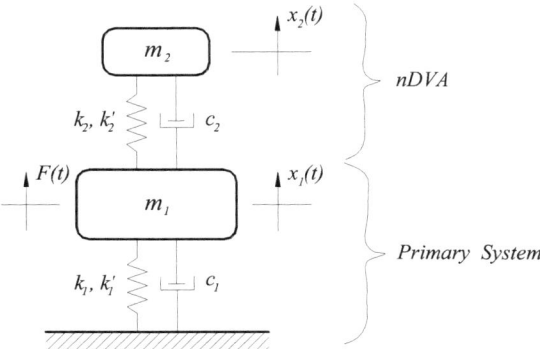

Figure 2.1 – Two degree-of-freedom nonlinear mechanical system.

The constitutive forces of the springs are taken into account as follows:

$$s_i(x_i) = k_i x_i + k_i' x_i^3, \quad i = 1, 2 \tag{2.1}$$

where x_1 and x_2 represent, respectively, the displacement of the primary system with respect to the ground and the displacement of the nDVA with respect to the primary mass. In the present case, the dampers are linear and the springs have nonlinear characteristics, where k_i and k_i' indicate, respectively, their linear and nonlinear coefficients. By applying the Newton's second law, the following equations of motion of the nonlinear system are obtained:

$$\begin{cases} (m_1 + m_2)\ddot{x}_1 + m_2\ddot{x}_2 + c_1\dot{x}_1 + k_1 x_1 = f_0 \cos(\omega t) - k_1' x_1^3 \\ m_2\ddot{x}_1 + m_2\ddot{x}_2 + c_2\dot{x}_2 + k_2 x_2 = -k_2' x_2^3 \end{cases} \tag{2.2}$$

Aiming at obtaining the dimensionless normalized equations of motion for the nonlinear dynamic system, the displacements are normalized according to the following relation $y_i = x_i/x_c$, where $x_c = f_0/k_1$ and $i=1,2$. In addition, the following expressions are introduced: $\zeta_1 = c_1/\left(2\sqrt{k_1 m_1}\right)$, $\zeta_2 = c_2/\left(2\sqrt{k_2 m_2}\right)$, $\varepsilon_1 = k_1^{nl} x_c^2/\left(m_1\omega^2\right)$, $\varepsilon_2 = k_2^{nl} x_c^2/\left(m_2\omega^2\right)$, $\mu = m_2/m_1$, $F = f_0/\left(m_1 x_c \omega^2\right)$, $\bar{\omega}_1^2 = k_1/m_1$, $\bar{\omega}_2^2 = k_2/m_2$, $\omega_1 = \bar{\omega}_1/\omega$, and $\omega_2 = \bar{\omega}_2/\omega$. After manipulations, the following matrix form of the normalized equations of motion is obtained:

$$\mathbf{M}\ddot{y}(t) + \mathbf{C}\dot{y}(t) + \mathbf{K}y(t) = F(t) \tag{2.3}$$

where the normalized mass, damping and stiffness matrices are expressed, respectively, as:

$$\mathbf{M} = \begin{bmatrix} 1+\mu & \mu \\ \mu & \mu \end{bmatrix}, \quad \mathbf{C} = \begin{bmatrix} 2\zeta_1\omega_1 & 0 \\ 0 & 2\mu\zeta_2\omega_2 \end{bmatrix}, \quad \mathbf{K} = \begin{bmatrix} \omega_1^2 & 0 \\ 0 & \mu\omega_2^2 \end{bmatrix} \tag{2.4}$$

and the normalized displacement and force vectors are given, respectively, as follows:

$$y(t) = \begin{Bmatrix} y_1(t) \\ y_2(t) \end{Bmatrix}, \quad F(t) = \begin{Bmatrix} F\cos(\omega t) - \varepsilon_1 y_1^3 \\ -\mu \varepsilon_2 y_2^3 \end{Bmatrix} \tag{2.5}$$

2.2.1 Steady-state harmonic responses of the nonlinear dynamic system

Various perturbation methods are based on averaging. This means that the unknown functions of the problem are considered as dependent variables by using a shift of variables from the original dependent variable (Nissen et. al., 1985; Nayfeh, 2000). These methods encompass techniques such as the following (Thomsen, 2003): Krylov-Bogoliubov method, Krylov-Bogoliubov-Mitropolsky method, and the method of the generalized average. In the present case, the Krylov-Bogoliubov method will be used to integrate Eq. (2.3), leading to an approximate solution of the nonlinear differential equation of motion. Within this context, the *Van der Pol Transformation* (Zhu et. al., 1992), represented by Eqs. (2.6) and (2.7), are used to guarantee that the transformation is unique.

$$y(\tau) = \mathbf{u}(\tau)\cos(\tau) + \mathbf{v}(\tau)\sin(\tau) \tag{2.6}$$

$$\dot{y}(\tau) = -\mathbf{u}(\tau)\sin(\tau) + \mathbf{v}(\tau)\cos(\tau) \tag{2.7}$$

where $\mathbf{u} = (u_1, u_2)^{\mathrm{T}}$ and $\mathbf{v} = (v_1, v_2)^{\mathrm{T}}$ are assumed to be slow functions of the normalized time $\tau = \omega t$.

After mathematical manipulation, we obtain a nonlinear algebraic system composed by four equations and four variables (u_1, u_2, v_1, v_2):

$$\begin{cases} \left(1 + \mu - \omega_1^2\right)u_1 + \mu u_2 - 2\zeta_1\omega_1 v_1 - \dfrac{3}{4}\varepsilon_1 u_1\left(u_1^2 + v_1^2\right) + F = 0 \\ \mu u_1 + \left(\mu - \mu\omega_2^2\right)u_2 - 2\mu\zeta_2\omega_2 v_2 - \dfrac{3}{4}\mu\varepsilon_2 u_2\left(u_2^2 + v_2^2\right) = 0 \\ \left(\omega_1^2 - 1 - \mu\right)v_1 - \mu v_2 - 2\zeta_1\omega_1 u_1 + \dfrac{3}{4}\varepsilon_1 v_1\left(u_1^2 + v_1^2\right) = 0 \\ \mu v_1 + \left(\mu - \mu\omega_2^2\right)v_2 + 2\mu\zeta_2\omega_2 u_2 - \dfrac{3}{4}\mu\varepsilon_2 v_2\left(u_2^2 + v_2^2\right) = 0 \end{cases} \tag{2.8}$$

Then, the obtained values of the parameters (u_1, u_2, v_1, v_2) are used to calculate the amplitudes of vibration of the nonlinear system, by considering the following relations $r_1 = \sqrt{u_1^2 + v_1^2}$ (for the primary mass) and $r_2 = \sqrt{u_2^2 + v_2^2}$ (for the nDVA). Moreover, for numerical computation, the following parameters of force and frequency are also considered: $\beta = F/\omega_1^2$ (force parameter) and $\Omega = \omega/\overline{\omega}_1$ (frequency parameter), so that in Eq. (2.8) one has $\omega_1 = 1/\Omega$ and $\omega_2 = \rho/\Omega$, where $\rho = \omega_2/\omega_1$ (frequency ratio).

2.3 SENSITIVITY ANALYSIS OF DYNAMIC RESPONSES

A mechanical system may exhibit physical and/or geometric nonlinear characteristics in the parameters of mass, stiffness and damping (Haug et al., 1986). Such functional dependence can be expressed in a general form as follows:

$$r = r\left(M(\mathrm{p}), C(\mathrm{p}), K(\mathrm{p})\right) \tag{2.9}$$

where r and p designate vectors of structural responses and design parameters, respectively.

The sensitivity of the structural responses with respect to a given parameter p_i, evaluated for a given set of values of the design parameter p_0 is defined as the following partial derivative:

$$\left.\frac{\partial r}{\partial \mathrm{p}_i}\right|_{\mathrm{p}^0} =$$
$$\lim_{\Delta \mathrm{p}_i \to 0}\left[\frac{r\left(M\left(\mathrm{p}_i^0 + \Delta \mathrm{p}_i\right), C\left(\mathrm{p}_i^0 + \Delta \mathrm{p}_i\right), K\left(\mathrm{p}_i^0 + \Delta \mathrm{p}_i\right)\right)}{\Delta \mathrm{p}_i} - \frac{r\left(M\left(\mathrm{p}_i^0\right), C\left(\mathrm{p}_i^0\right), K\left(\mathrm{p}_i^0\right)\right)}{\Delta \mathrm{p}_i}\right] \tag{2.10}$$

where $\Delta \mathrm{p}_i$ is an arbitrary variation applied to the current value of parameter p_i^0, while all other parameters remain unchanged. The sensitivity with respect to p_i can be estimated by finite differences, thus computing successively the responses corresponding to $\mathrm{p}_i = \mathrm{p}_i^0$ and $\mathrm{p}_i = \mathrm{p}_i^0 + \Delta \mathrm{p}_i$.

Such procedure is an estimated approach enabling to calculate the sensitivity of the dynamic system responses with respect to small modifications introduced in the design parameters. Moreover, the results depend upon the choice of the value of the parameter increment $\Delta \mathrm{p}_i$. Another strategy consists in computing the analytical derivatives, if possible, of the structural responses with respect to the parameters of interest.

To illustrate the computation procedure for the sensitivity of dynamic responses, numerical tests were performed by using the system configuration illustrated in Fig. 2.1. As previously mentioned, the computations are devoted to obtaining the sensitivities of the driving point frequency responses, which are given by the elements of $H(\omega, \mathrm{p})$.

In this example, the normalized structural parameters ζ_1, ζ_2, ε_1, ε_2, β, $\varOmega$, and ρ are considered as the design variables in the computation of the normalized sensitivities of the frequency responses with respect to a given parameter p, $S(\omega, \mathrm{p})$.

The computation consist in obtaining the driving point dynamic responses $H(\omega, p)$ associated to the displacement x_1 indicated in Fig. 2.1, in the frequency band of interest, i.e., Ω=[0.7–1.3 Hz], comprising a total number of 300 frequency points. Moreover, for the numerical resolution of the Eq. (2.8) the following initial conditions have been adopted (u_1=5, u_2=5, v_1=2 and v_2=4) by using the Newton Method to solve the nonlinear equations system.

The normalized real parts of the approximated complex sensitivity functions calculated by finite differences are shown in Figs. 2.2 to 2.5, for which a variation of 20% of the nominal values of each design parameter was adopted (ε_1=0.001, ε_2=0.01, β=0.1, ζ_1=0.01, ζ_2=0.01, $\boxed{}$=0.05 and ρ=). In these figures, the real parts of the frequency responses $H(\omega,p)$ were multiplied by convenient scale factors (SF). The sensitivity functions, denoted by $S(\omega,p)$, have been normalized according to the following scheme:

$$S(\omega,p) = \left.\frac{\partial S}{\partial p}\right|_{(\omega,p_0)} \frac{p_0}{H(\omega,p)} \qquad (2.11)$$

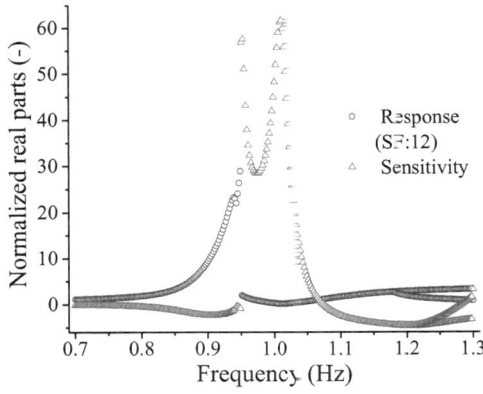

Figure 2.2 – Sensitivity of $H(\omega,p)$ with respect to ρ.

13

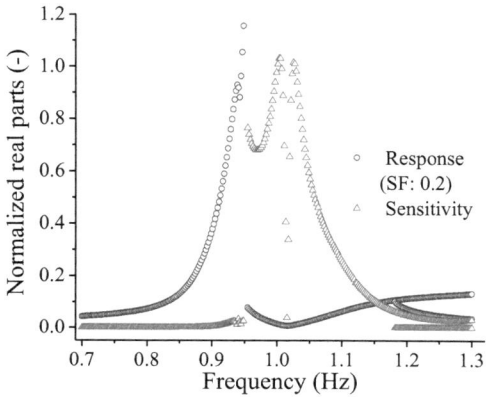

Figure 2.3 – Sensitivity of H(ω,p) with respect to ε₂.

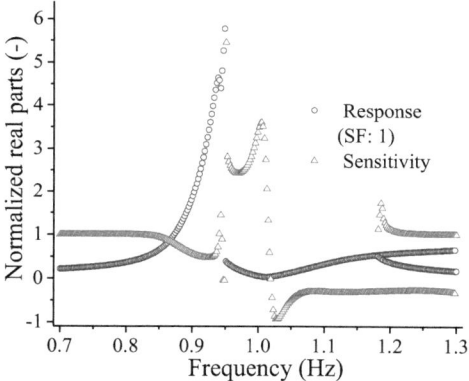

Figure 2.4 – Sensitivity of H(ω,p) with respect to β.

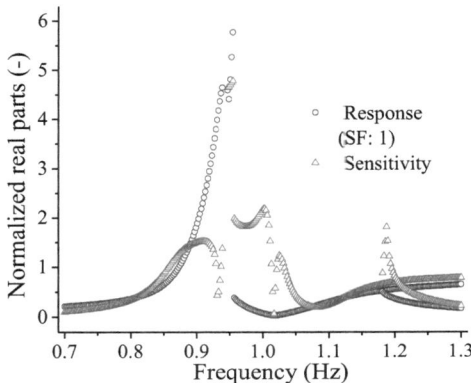

Figure 2.5 – Sensitivity of H(ω,p) with respect to β

Based on the amplitudes and signs of the sensitivity functions, one can evaluate the degree of influence of the design variables upon the suppression bandwidth, in the frequency band of interest. In this context, the parameters ζ_1, ζ_2 and ε_1 do not have a significant influence on the evaluation of the suppression bandwidth. Consequently, these parameters are not considered as design variables in the optimization run, as properly described in the next chapter.

2.4 SIMULATION OF nDVA BASED ON SENSITIVITY ANALYSIS

After having verified the influence of each design variable on the dynamic response of the nonlinear system, the interest now is to simulate the nDVA considering different configurations for the following parameters: ε_2, β, β and ρ. The other parameters are considered as constants (ε_1=0.001, ζ_1=0.01 and ζ_2=0.01), Borges (2010).

2.4.1 Nonlinear Dynamic Vibration Absorber (ρ=1, μ=0.05, β=0.1 and ε_2=0.01)

Figure 2.6 presents a simulation of the forcing frequency versus the displacement amplitude for the linear and nonlinear dynamic vibration absorbers considering the following parameters: ρ=1, μ=0.05, β=0.1 and ε_2=0.01.

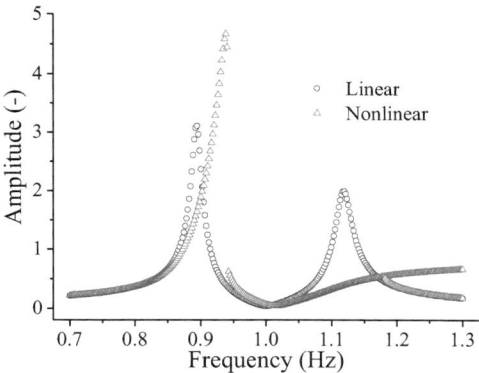

Figure 2.6 – Forcing frequency versus displacement amplitude considering the linear and nonlinear dynamic vibration absorber (ρ=1, μ=0.05, β=0.1 and ε_2=0.01).

Looking at the non-linear response, it is noted that there arises a region of instability, which comprises the frequency band ranging from Ω=0.91 to Ω=0.94. An interesting result is that the use of the absorber considering the nonlinear spring enables a considerable attenuation of the vibration amplitude for Ω >1.

2.4.2 Nonlinear Dynamic Vibration Absorber (ρ=1.2, μ=0.03, β=0.08 and ε_2=0.02)

Figure 2.7 presents a simulation of the forcing frequency versus the displacement amplitude for a linear and a nonlinear dynamic vibration system considering the following parameters: ρ=1.2, μ=0.03, β=0.08 and ε_2=0.02. In this case there was a small increase in the values of the parameters ρ (frequency ratio) and ε_2 (nonlinear stiffness coefficient of the spring that connects the primary mass to the secondary mass). At the same time as to reduce μ values (ratio of masses) and β (force parameter).

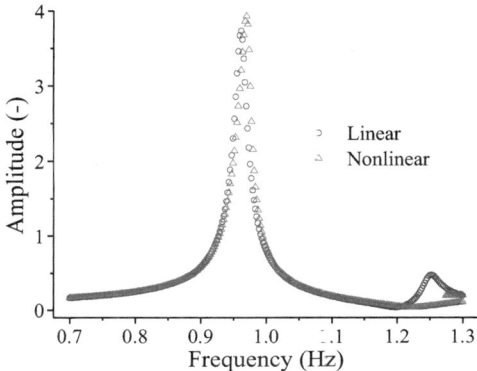

Figure 2.7 – Forcing frequency versus displacement amplitude considering the linear and nonlinear dynamic vibration absorber ($\rho=1.2$, $\mu=0.03$, $\beta=0.08$ and $\varepsilon_2=0.02$).

It can be seen that these variables directly influence the system response. In comparison with Fig. 2.6, shown above, there was a significant reduction in the amplitude of vibration of the system. Besides, the nonlinear system instability region corresponds to $\Omega>1.25$.

Another interesting point is that, despite the increase of the nonlinear parameters lead to undesired instabilities in the system, they are also very helpful to attenuate the vibration of the system.

2.4.3 Nonlinear Dynamic Vibration Absorber ($\rho=0.8$, $\mu=0.08$, $\beta=0.12$ and $\varepsilon_2=0.009$)

Finally, a simulation of the forcing frequency versus the displacement amplitude for linear and nonlinear dynamic vibration absorbers considering the parameters $\rho=0.8$, $\mu=0.08$, $\beta=0.12$ and $\varepsilon_2=0.009$ is presented in Fig. 2.8.

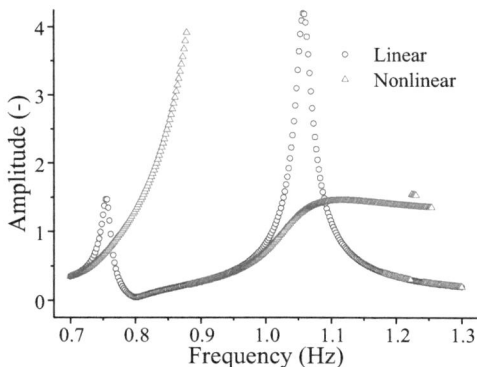

Figure 2.8 – Forcing frequency versus displacement amplitude considering the linear and nonlinear dynamic vibration absorbers (ρ=0.8, μ=0.08, β=0.12 and ε_2=0.009).

In this case, the values of the parameters ρ and ε_2 are smaller than those considered in the case 1 (see Fig. 2.6) and the values of the parameters μ and β are greater than those considered in the case 1 (see Fig. 2.6). It becomes evident that beyond of ε_2, other parameters contributes to the appearance of instabilities in the response, thus, increasing other parameters that are also fairly significant in system behavior, we can see a nonlinear behavior and it can be seen instabilities regions ranging up to near Ω=0.8 until Ω=0.9.

One can also see that decreasing the value of ε_2 results an increased amplitude of vibration of the system, thus confirming that this factor, despite bringing instability to the system, is very important to attenuate the vibration response.

2.5 PRELIMINARY CONCLUSIONS

In this chapter the simulation and sensitivity analysis of a nonlinear dynamic vibration absorber was performed. The system nonlinearity was introduced in the springs that connect the primary mass to the ground and the absorber to the primary mass, respectively. The choice of the design variables that should be used in next chapter was based on the sensitivity analysis performed in this chapter. It is worth mentioning that these parameters are directly associated with the effectiveness of the nDVA (perspective of vibration reduction).

In terms of the system resolution, the equations of motion of the nonlinear two d.o.f. system were numerically integrated by using the so-called average method that provides an approximate solution to nonlinear dynamic problems. The nonlinear algebraic equations obtained were solved numerically, enabling the determination of the roots of the nonlinear algebraic equations. It is worth mentioning that the nonlinearity factor is an important parameter to be investigated during the design procedure of nDVAs, due to its contribution to the reduction of the vibration level. However, care must be taken with high values of nonlinearity since these values lead to unstable conditions for the system.

CHAPTER 3:
DESIGN OF A NONLINEAR DYNAMIC VIBRATION ABSORBER USING BIO-INSPIRED OPTIMIZATION METHODS

3.1 INTRODUCTION

Nowadays, biological systems have contributed significantly to the development of new optimization techniques. In this way, optimization algorithms mimic the behavior of various species found in the nature to update a population of candidates (Lobato et. al., 2010; 2012). In addition, this capacity makes possible the communication among the agents (individuals of the population) that capture the changes in the environment generated by local interactions (Parrish et. al., 2002). These techniques, known as Bio-inspired Optimization Methods (BiOM), are able to notice and modify the environment in order to seek for diversity and convergence (fundamental requirements to solve optimization problems).

In this chapter, a review of three BiOM algorithms (Bees Colony Algorithm – BCA, Firefly Colony Algorithm - FCA and Fish Swarm Algorithm - FSA) is presented. Moreover, the basic concepts of multi-objective optimization and the extension of BiOM to the multi-objective context is addressed. Finally, the design of a nonlinear dynamic vibration absorber considering these BiOM both in the mono and in multi-objective contexts is performed.

3.2. BIOM REVIEW
3.2.1 Bees Colony Algorithm

The Bees Colony Algorithm (BCA) is based on the behavior of a colony of honey bees observed in nature. The colony can extend itself over long distances and in multiple directions simultaneously to exploit a large number of food sources. In addition, the colony of honey bees presents as characteristic the capacity of memorization, learning and transmission of information, thus forming the so-called swarm intelligence (von Frisch, 1976).

In a colony, the foraging process begins by scout bees being sent to search randomly for promising flower patches. When they return to the hive, those scout bees that found a patch that is rated above a certain quality threshold (measured as a combination of some constituents, such as sugar content) deposit their nectar or pollen and go to the waggle dance. This dance is responsible for the transmission (colony communication) of information regarding a particular flower patch: the direction in which it can be found, its distance from the hive and its quality rating (or fitness) (von Frisch, 1976). The waggle dance enables the colony to evaluate the relative merit of different patches according to both the quality of the food they provide and the amount of energy needed to harvest it (Camazine et. al., 2003). After waggle dancing, the dancer (scout bee) goes back to the flower patch with follower bees that were waiting inside the hive. More follower bees are sent to more promising patches. This allows the colony to gather food quickly and efficiently. While harvesting from a patch, the bees monitor its food level. This is necessary to decide upon the next waggle dance when they return to the hive (Camazine et. al., 2003). If the patch is still good enough as a food source, then it will be advertised in the waggle dance and more bees will be recruited to that source. In this context, Pham et al. (2006) proposed an optimization algorithm inspired by the natural foraging behavior of honey bees (Bees Colony Algorithm - BCA) as presented in Tab. 3.1.

Table 3.1 – Basic step in the Bees Colony Algorithm (Pham et. al., 2006).

1-Initialize population with random solutions
2-Evaluate fitness of the population
3-While (stopping criterion not met)
4-Select sites for neighborhood search
5-Recruit bees for selected sites and evaluate fitnesses
6-Select the fittest bee from each site
7-Assign remaining bees to search randomly and evaluate their fitnesses
8-End while

The BCA requires a number of parameters to be set, namely, the number of scout bees (n), number of sites selected for neighborhood search (out of n visited sites) (m), number of top-rated (elite) sites among m selected sites (e), number of bees recruited for the best e sites (n_{ep}), number of bees recruited for the other ($m-e$) selected sites, neighborhood search (ngh), and the stopping criterion.

The BCA starts with the n scout bees being placed randomly in the search space. The fitnesses of the sites visited by the scout bees are evaluated in step 2. In step 4, bees that have the highest fitnesses are chosen as selected bees and sites visited by them are chosen for neighborhood search. Then, in steps 5 and 6, the algorithm conducts searches in the neighborhood of the selected sites, assigning more bees to search near to the best *e* sites. The bees can be chosen directly according to the fitnesses associated with the sites they are visiting.

Alternatively, the fitness values are used to determine the probability of the bees being selected. Searches in the neighborhood of the best *e* sites, which represent more promising solutions, are made more detailed by recruiting more bees to follow them than the other selected bees. Together with scouting, this differential recruitment is a key operation of the BCA. However, in step 6, for each patch only the bee with the highest fitness will be selected to form the next bee population. In nature, there is no such restriction. This restriction is introduced here to reduce the number of points to be explored. In step 7, the remaining bees in the population are assigned randomly around the search space scouting for new potential solutions.

In the literature, various applications using this bio-inspired approach can be found, such as: modeling combinatorial optimization transportation engineering problems (Lucic and Teodorovic, 2001), optimal control problems (Afshar et. al., 2001), engineering system design (Yang, 2005; Lobato et. al., 2010), mathematical function optimization (Pham et. al., 2006), transport problems (Teodorovic and Dell'Orco, 2005), dynamic optimization (Chang, 2006), parameter estimation in control problems (Azeem and Saad, 2004; Lobato et. al., 2012), design of a nonlinear mechanical system (Borges et. al., 2013), machinability of stainless steel using multi-objective optimization and BiOM (Lobato et. al., 2014), development of a a bio-inspired energy-efficient clustering protocol for mobile learning (Xia et. al., 2014), parameter estimation for crop growth model (Zúniga et al., 2014), among other applications (http://www.bees-algorithm.com/).

3.2.2 Firefly Colony Algorithm

The Firefly Colony Algorithm (FCA) is based on the characteristic of the bioluminescence of fireflies, insects notorious for their light emission. According to Yang (2008), biology does not have a complete knowledge to determine all the utilities that firefly luminescence can bring to, but at least three functions have been identified: *i*) as a communication tool and appeal to potential partners in reproduction, *ii*) as a bait to lure potential prey for the firefly, *iii*) as a warning mechanism for potential predators reminding them that fireflies have a bitter taste.

In this way, the bioluminescent signals are known to serve as elements of courtship rituals (in most cases, the females are attracted by the light emitted by the males), methods of prey attraction, social orientation or as a warning signal to predators (Lukasik and Zak, 2009).

Some of the flashing characteristics of fireflies were idealized to develop firefly-inspired algorithms. For simplicity the following three idealized rules are used (Yang, 2010):

- all fireflies are unisex, i.e., one firefly will be attracted to other fireflies regardless of their gender;
- attractiveness is proportional to their brightness; thus, for any two flashing fireflies, the less brighter will move towards the brighter one. The attractiveness is proportional to the brightness and they both decrease as their distance increases. If there is no brighter one than a particular firefly, it will move randomly;
- the brightness of a firefly is affected or determined by the landscape of the objective function. For a maximization problem, the brightness can simply be proportional to the value of the objective function.

According to Yang (2008), in the firefly algorithm there are two important issues: the variation of light intensity and the formulation of the attractiveness. For simplicity, it is always assumed that the attractiveness of a firefly is determined by its brightness, which in turn is associated with the encoded objective function.

This swarm intelligence optimization technique is based on the assumption that the solution of an optimization problem can be perceived as agent (firefly) which glows proportionally to its quality in a considered problem setting. Consequently, each brighter firefly attracts its partners (regardless of their gender), which makes the search space being explored more efficiently. The algorithm makes use of a synergic local search. Each member of the swarm explores the problem space taking into account results obtained by others, still applying its own randomized moves as well. The influence of other solutions is controlled by the attractiveness value (Lukasik and Zak, 2009).

According to Lukasik and Zak (2009), the FCA can be presented as follows. Consider a continuous constrained optimization problem where the task is to minimize the cost function $f(x)$.

$$f\left(x^*\right)= \min_{x \in S} f\left(x\right) \qquad (3.1)$$

Assume that there exists a swarm of agents (fireflies) solving the above mentioned problem iteratively and x_i represents a solution for a firefly i in algorithm's iteration k, whereas $f(x_i)$ denotes its cost. Initially, all fireflies are dislocated in S (randomly or employing some deterministic strategy). Each firefly has its distinctive attractiveness λ, which implies how strong it attracts other members of the swarm. As the firefly attractiveness, one should select any monotonically decreasing function of the distance $r_j=d(x_i,x_j)$ to the chosen firefly j, i.e., the exponential function:

$$\lambda = \lambda_0\exp\left(-\gamma r_j\right) \tag{3.2}$$

where λ_0 and γ are predetermined algorithm parameters, namely the maximum attractiveness value and the absorption coefficient, respectively. Furthermore, every member of the swarm is characterized by its light intensity (I_i) which can be directly expressed as an inverse of a cost function $f(x_i)$. To effectively explore considered search space S it is assumed that each firefly i is changing its position iteratively taking into account two factors: attractiveness of other swarm members with higher light intensity, $I_j > I_i$, for all $j = 1, ..., m$, $j \neq i$ which is varying across distance and a fixed random step vector u_i. It should be noted as well that if no brighter firefly can be found only such randomized step is used.

Thus, moving at a given time step t of a firefly i toward a better firefly j is defined as:

$$x_i^t = x_i^{t-1} + \lambda\left(x_j^{t-1} - x_i^{t-1}\right) + \alpha\left(rand - 0.5\right) \tag{3.3}$$

where the second term on the right hand side of the equation inserts the factor of attractiveness λ while the third term, governed by α parameter, governs the insertion of certain randomness in the path followed by the firefly, *rand* is a random number between 0 and 1.

In the literature, few works using the FCA can be found. In this context, the interest is focused on applications dedicated to continuous constrained optimization (Lukasik and Zak, 2009), multimodal optimization (Yang, 2009), solution of singular optimal control problems (Pfeifer and Lobato, 2010), economic emissions load dispatch problem (Apostolopoulos and Vlachos, 2011), parameter estimation in control problems (Lobato et. al., 2012), design of a nonlinear mechanical system (Borges et. al., 2013), machinability of stainless steel using multi-objective optimization and BiOM (Lobato et. al., 2014), among other applications.

3.2.3 Fish Swarm Algorithm

The Fish Swarm Algorithm (FSA) is based on fish swarm as observed in nature. The following characteristics are considered (Li et. al., 2002; Madeiro, 2010): *i*) each fish represents a candidate solution of the optimization problem; *ii*) food density is related to an objective function to be optimized (in an optimization problem, the amount of food in a region is inversely proportional to the value of the objective function); and *iii*) the aquarium is the design space where the fish is found.

As noted earlier, the fish weight as the swarm represents the accumulation of food (i.e., the objective function) received during the evolutionary process. In this case, the weight is an indicator of success (Li et. al., 2002; Madeiro, 2010). Basically, the FSA presents four operators classified into two classes, i.e., "food search" and "movement". Details on each of these operators are shown in the following.

Individual Movement Operator

This operator contributes to the individual and collective movement of fishes in the swarm. Each fish updates its new position using Eq. (3.4):

$$x_i^{t+1} = x_i^t + rand \times s_{ind} \qquad (3.4)$$

where x_i is the final position of fish i at current generation, *rand* is a random generator and s_{ind} is a weighted parameter.

Food Operator

The weight of each fish is a metaphor used to measure the success of food search. The higher the weight of a fish, the more likely this fish is in a potentially interesting region in the design space.

According to Madeiro (2010), the amount of food that a fish eats depends on the improvement of the objective function in the current iteration. The weight is updated according to Eq. (3.5):

$$W_i^{t+1} = W_i^t + \frac{\Delta f_i}{max(\Delta f)} \qquad (3.5)$$

where W_i^t is the fish weight i at generation t and Δf_i is the difference of the objective function between the current position and the new position of fish i. It is important to emphasize that $\Delta f_i = 0$ for the fishes in same position.

Instinctive Collective Movement Operator

This operator is important for the individual movement of fishes when $\Delta f_i \neq 0$. Thus, only the fishes whose individual execution of the movement resulted in improvement of their fitness will influence the direction of motion of the school, resulting in instinctive collective movement. In this case, the resulting direction (I), calculated using the contribution of the directions taken by the fish, and the new position of the i-th fish are given by:

$$\vec{I}^t = \frac{\sum_{i=1}^{N} \Delta \vec{x}_i \Delta f_i}{\sum_{i=1}^{N} \Delta f_i} \tag{3.6}$$

$$\vec{x}_i^{t+1} = \vec{x}_i^t + \vec{I}^t \tag{3.7}$$

It is worth mentioning that in the application of this operator the direction chosen by a fish that was able to locate the largest portion of food exerts the greatest influence on the swarm. Therefore, the instinctive collective movement operator tends to guide the swarm in the direction of motion chosen by that particular fish (the one that found the largest portion of food in its individual movement).

Non-Instinctive Collective Movement Operator

As noted earlier, the fish weight is a good indication of search success for food. In this way, if the swarm weight is increasing, it means that the search process is successful. So, the "radius'" of the swarm must decrease so that other regions can be explored. Otherwise, if the swarm weight remains constant, the radius should increase to allow the exploration of new regions.

For the swarm contraction, the centroid concept is used. This is done by means of an average position of all fishes weighted by their respective fish weights, according to Eq. (3.8):

$$\vec{B}^t = \frac{\sum_{i=1}^{N} \vec{x}_i W_i^t}{\sum_{i=1}^{N} W_i^t} \tag{3.8}$$

If the swarm weight remains constant in the current iteration, all fish must update their positions by using Eq. (3.9):

$$\vec{x}^{t+1} = \vec{x}^t - s_{vol} rand \frac{\vec{x}^t - \vec{B}^t}{d\left(\vec{x}^t, \vec{B}^t\right)} \tag{3.9}$$

where d is a function that calculates the Euclidean distance between the centroid and the current fish position, and s_{vol} is the step size used to control fish displacements.

In the literature, few works using the FSA can be found. In this context, the following can be cited: parameter estimation in engineering systems (Li et al., 2004), feed forward neural networks (Wang et al., 2005), combinatorial optimization (Cai, 2010), Augmented Lagrangian fish swarm based method for global optimization (Rocha et. al., 2011), forecasting stock indices using radial basis function neural networks optimized (Shen et al., 2011), hybridization of the FSA with the Particle Swarm Algorithm to solve engineering systems (Tsai and Lin, 2011), parameter estimation in control problems (Lobato et al., 2012), design of a nonlinear mechanical system (Borges et. al., 2013), machinability of stainless steel using multi-objective optimization and BiOM (Lobato et al., 2014), among other applications.

3.3 MULTI-OBJECTIVE OPTIMIZATION

When dealing with Multi-objective Optimization Problems (MOP), the notion of optimality has to be extended. The most common one in the current literature is the one originally proposed by Edgeworth (1881) and later generalized by Pareto (1971). This notion is called Edgeworth-Pareto optimality, or simply Pareto optimality, and refers to finding good tradeoffs among all the objectives. This definition leads to a set of solutions that is called the Pareto optimal set, whose corresponding elements are called non-dominated or non-inferior. The concept of optimality for single objective optimization is not directly applicable in MOPs. For this reason, a classification of the solutions is introduced in terms of Pareto optimality, according to the following definitions (Deb, 2001):

Definition 3.1 - The Multi-objective Optimization Problem (MOP) can be defined as:

$$f(x) = (f_1(x), f_2(x), ..., f_m(x)), \ m = 1, ..., M \qquad (3.10)$$

subject to

$$h(x) = (h_1(x), h_2(x), ..., h_i(x)), \ i = 1, ..., H \qquad (3.12)$$

$$g(x) = (g_1(x), g_2(x), ..., g_j(x)), \ j = 1, ..., J \qquad (3.13)$$

$$x = (x_1, x_2, ..., x_n), \ n = 1, ..., N, \ x \in X \qquad (3.14)$$

where x is the vector of design (or decision) variables, f is the vector of objective functions and X is denoted as the design (or decision) space. The constraints h and g (≥ 0) determine the feasible region.

Definition 3.2 - Pareto Optimality: when the set P is the entire search space, or P = S, the resulting non-dominated set P' is called the Pareto-optimal set. Similar to global and local optimal solutions in the case of single-objective optimization, global and local Pareto-optimal sets appear in multi-objective optimization.

In the multi-objective context, various Multiple-Objective Evolutionary Algorithms (MOEAs) are appropriate to handle the associate problems. This group of algorithms conjugates the basic concepts of dominance described in the latter section with the general characteristics of evolutionary algorithms. MOEAs are able to deal with non-continuous, non-convex and/or non-linear spaces, as well as problems whose objective functions are not explicitly known (Deb, 2001). Basically, the main features of these MOEAs are the following:

- Mechanism of adaptation assignment in terms of dominance: between one non-dominated solution and a dominated one, the algorithm will favor the non-dominated. Moreover, when both solutions are equivalent in dominance, the one located in a less crowded area will be favored. Finally, the extreme points (i.e., the solutions that have the best value with respect to a particular objective) of the non-dominated population are preserved and their adaptation is better than any other non-dominated point, so that maximum front expansion is allowed.

- Incorporation of elitism: the elitism is commonly implemented by using a secondary population of non-dominated solutions previously stored. When performing recombination (selection-crossover-mutation), parents are taken from this file in order to produce the offspring.

3.4 BIO-INSPIRED MULTI-OBJECTIVE OPTIMIZATION

Due to the success obtained by BiOM in different applications in science and engineering, literature presents a number of examples involving the use of BiOM to solve multi-objective optimization problems. In this work, the MOBC (Multi-objective Optimization Bee Colony), MOFC (Multi-objective Optimization Firefly Colony) and MOFS (Multi-objective Optimization Fish Swarm) algorithms are used. Each proposed algorithm (BCA, FCA and FSA) is coded according to the following structure (see Fig. 3.1): an initial population of size N is randomly generated (considering each evolutionary algorithm). All dominated solutions are removed from the population through the operator Fast Non-Dominated Sorting. In this way, the population is sorted into non-dominated fronts F_j (sets of vectors that are non-dominated with respect to each other). This procedure is repeated until each vector is member of a front. A child is generated from the three parents (this process continues until N children are generated). The new population is then classified according to the dominance criterion. If the number of individuals of this population is larger than a number defined by the user, it is truncated obeying the criterion named the Crowding Distance (Deb, 2001). The Crowding Distance describes the density of solutions surrounding a vector. To compute the Crowding Distance for a set of population members, the vectors are sorted according to their objective function value for each

Objective Function. To the vectors with the smallest or largest values an infinite Crowding Distance (or an arbitrarily large number for practical purposes) is assigned. For all other vectors the Crowding Distance is calculated according to:

$$dist_{x_i} = \sum_{j=0}^{m-1} \frac{f_{j,i+1} - f_{j,i-1}}{|f_{j,max} - f_{j,min}|} \tag{3.15}$$

where f_j corresponds to the j-th objective function and m equals the number of objective functions.

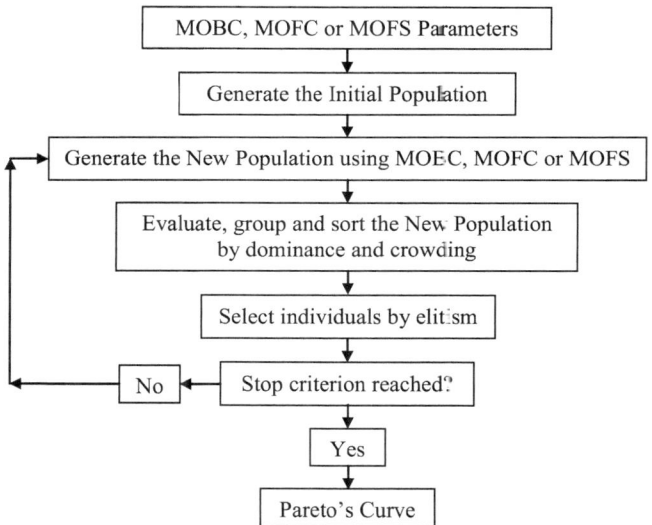

Figure 3.1 – Flowchart BIOM.

3.5 OPTIMAL DESIGN OF A NONLINEAR DYNAMIC VIBRATION ABSORBER

Now the optimal design of nonlinear optimization problem devoted to a two-degree-of-freedom nonlinear damped system constituted of a primary mass attached to the ground by a linear spring and a secondary mass attached to the primary system by a nonlinear spring (see Fig. 1.1 in Chapter 2) by using BiOM is considered. Previously, the influence of each design variable on the dynamic response of the nonlinear system was verified. As observed in Figs. 2.2 to 2.5 (Chapter 2), the degrees of influence of the parameters ε_2, μ, β and ρ on the suppression bandwidth and response amplitude are significant. For this reason, these parameters will be considered as design variables in the optimization process.

3.5.1 Mono-objective Optimization of a Nonlinear Dynamic Vibration Absorber

This section is dedicated to presenting alternative techniques for the optimal design of a nonlinear dynamic vibration absorber (nDVA) by using BiOM.

For design purposes the following steps are established:

- Objective function: maximization of the bandwidth, as showed in Fig. 3.2;

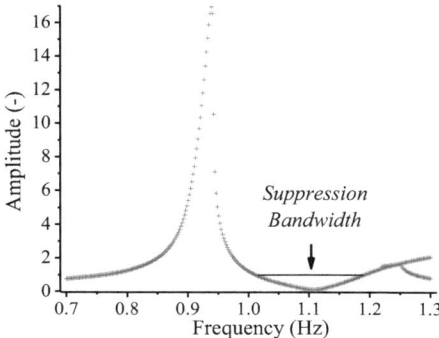

Figure 3.2 – Representation of the suppression bandwidth.

- Design variables (normalized structural parameters): $0.9 \leq \rho \leq 1.2$, $0.04 \leq \mu \leq 0.06$, $0.09 \leq \beta \leq 0.12$ and $0.009 \leq \varepsilon_2 \leq 0.012$;
- The parameters used by the BiOM: BCA - number of scout bees (5), number of bees recruited for the best e sites (5), number of bees recruited for the other selected sites (5), number of sites selected for neighborhood search (5), number of top-rated (elite) sites among m selected sites (5), neighborhood search (ngh) (10^{-3}) and generation number (100); FCA - number of fireflies (25), maximum attractiveness value (0.9), absorption coefficient (0.7) and generation number (100); FSA - number of fishes (25), weighted parameter value (1), control fish displacements (10^{-1}) and generation number (100);
- In order to examine the considered quality of the solution methodology, the Genetic Algorithm (GA) and the Particle Swarm Optimization (PSO) techniques were performed. The parameters used by GA and PSO are as follows: GA parameters - population size (25), crossover rate (0.8), mutation rate (0.01), and generation number (100); PSO parameters - population size (25), inertia weight (1), cognitive and social parameters (0.5), constriction factor (0.8), and generation number (100);
- For the considered parameters, the number of objective function evaluations is

25+25×100;

- Stopping criterion: a given number of generations is defined to interrupt the procedure;
- Each algorithm was run 20 times by using 20 different seeds for the random generation of the initial population.
- To solve the nonlinear equations system, the Newton Method is used.

In Table 3.2 the results (best (B), average (A) and worst (W)) obtained for the design of the nonlinear vibration absorber are presented.

Table 3.2 – Results obtained to maximize the bandwidth by using the BIOM.

Strategy		ρ	ε_1	β	ε_2	Bandwidth
BCA	B	1.165	0.053	0.110	0.017	0.236
	A	1.189	0.055	0.102	0.015	0.230
	W	1.027	0.058	0.091	0.011	0.218
FCA	B	1.158	0.054	0.110	0.017	0.230
	A	1.158	0.054	0.110	0.017	0.230
	W	1.157	0.053	0.110	0.017	0.228
FSA	B	1.166	0.052	0.109	0.016	0.235
	A	1.191	0.050	0.101	0.016	0.232
	W	1.019	0.053	0.092	0.010	0.219
GA	B	1.101	0.054	0.099	0.098	0.233
	A	1.136	0.052	0.092	0.012	0.226
	W	0.932	0.044	0.110	0.012	0.228
PSO	B	1.101	0.053	0.099	0.099	0.233
	A	1.132	0.051	0.095	0.010	0.224
	W	0.918	0.049	0.103	0.011	0.225

This table shows that the BiOM algorithms presented good estimates for the unknown parameters as compared with the results from GA and PSO.

3.5.2 Multi-objective Optimization of a Nonlinear Dynamic Vibration Absorber

In this section the optimal design of a nonlinear dynamic vibration absorber (nDVA) by using BiOM is presented. For design purposes, the following steps are established:

- The deterministic optimization problem is composed by the two following objective functions: the first cost function is the amplitude of the dynamic response of the nonlinear damped system corresponding to the mode #1 (by minimizing the amplitude of the response at the corresponding resonance peak); the second cost function is the maximization of the suppression bandwidth, as showed in Fig. 3.3;

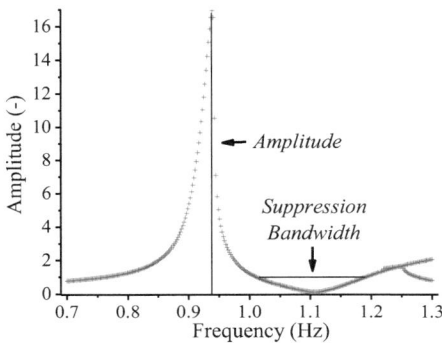

Figure 3.3 – Representation of suppression bandwidth and amplitude.

- Design variables (normalized structural parameters): $0.8 \leq \rho \leq 1.2$, $0.045 \leq \mu \leq 0.08$, $0.08 \leq \beta \leq 0.12$ and $0.009 \leq \varepsilon_2 \leq 0.012$;
- The parameters used by the BiOM: MOBC - number of scout bees (5), number of bees recruited for the best e sites (5), number of bees recruited for the other selected sites (5), number of sites selected for neighborhood search (3), number of top-rated (elite) sites among m selected sites (2), neighborhood search (ngh) (10^{-3}) and generation number (100); MOFC - number of fireflies (20), maximum attractiveness value (0.9), absorption coefficient (0.7) and generation number (100); MOFS - number of fishes (20), weighted parameter value (1), control fish displacements (10^{-1}) and generation number (100);
- The computation consists in obtaining the driving point dynamic responses $H(\omega, p)$ associated to the displacement x_1 as indicated on Fig. 2.1 (see Chapter 2), in the frequency band of interest $\Omega=[0.7–1.3$ Hz] comprising a total number of 300 frequency points. For the numerical solution of Eq. (2.8) (see Chapter 2) the Newton Method was used, for which the following initial conditions were adopted ($u_1=5$, $u_2=5$, $v_1=2$ and $v_2=4$);
- For comparison purposes the results obtained from MOBC, MOFC and MOFS, the NSGA II – *Nondominated Sorting Genetic Algorithm* (Deb et. al., 2000)

will be used. The parameters of the NSGA II are defined as follows: probability of selection, 0.25; probability of crossover, 0.25; probability of mutation, 0.25; number of generations, 100; number of individuals per generation, 20; and the sharing coefficient, 0.2. For the considered parameters, the number of objective function evaluations is $20+20\times100$;

- Stopping criterion: a given number of generations is defined to interrupt the procedure;
- Each algorithm was run 10 times by using 10 different seeds for the random generation of the initial population.

Figure 3.4 present the Pareto's Curve considering the BiOM strategies.

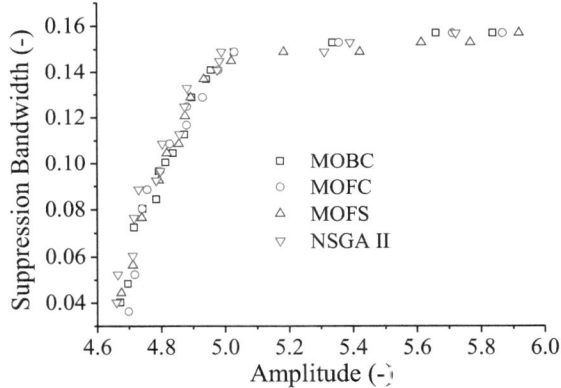

Figure 3.4 – Pareto's Curve by using the BIOM strategies.

In this figure it is possible to observe that the MOBC, MOFC, MOFS and NSGA II were able to estimate satisfactory the Pareto's Curve, resulting a similar number of objective function evaluations for all the techniques conveyed. In addition, it is important to observe the conflicting behavior of the two objectives, i.e., the "best" value in terms of the maximization of the bandwidth results in a "worst" value in terms of the minimization of the amplitude. In the other hand, the "best" value in terms of the minimization of the amplitude results in a "worst" value in terms of the maximization of the bandwidth.

Table 3.3 present selected points ("best" value in terms of the maximization of the bandwidth and "best" value in terms of the minimization of the amplitude) of the Pareto's Curve obtained by the BiOM.

Table 3.3 – Selected points (Extremes) of the Pareto's Curve by using the BIOM.

Strategy	ρ	μ	β	ε_2	Amplitude	Bandwidth
MOBC	9.03E-01	5.45E-02	9.00E-02	9.41E-03	4.67	4.02E-02
	1.09	5.29E-02	1.07E-01	1.09E-02	5.83	1.57E-01
MOFC	9.00E-01	5.41E-02	9.00E-02	9.32E-03	4.69	3.62E-02
	1.09	5.41E-02	1.07E-01	1.09E-02	5.86	1.57E-01
MOFS	9.05E-01	5.45E-02	9.00E-02	9.14E-03	4.67	4.42E-02
	1.09	5.32E-02	1.09E-01	1.04E-02	5.91	1.57E-01
NSGA II	9.00E-01	5.42E-02	9.00E-02	9.25E-03	4.66	4.03E-02
	1.09	5.29E-02	1.06E-01	1.09E-02	5.73	1.58E-01

As observed in Fig. 3.4, the results obtained by using the MOBC, MOFC, MOFS and NSGA II were very similar. Figures 3.5 and 3.6 present the simulation of nDVA considering the following parameters obtained by using the BiOM (case a – best value considering the amplitude minimization; and case b – best value considering the bandwidth maximization):

- MOBC: case a (ρ=9.03E-01, μ=5.45E-02, β=9.00E-02 and ε_2=9.41E-03) and case b (ρ=1.09, μ=5.29E-02, β=1.07E-01 and ε_2=1.09E-02);
- MOFC: case a (ρ=9.00E-01, μ=5.41E-02, β=9.00E-02 and ε_2=9.32E-03) and case b (ρ=1.09, μ=5.41E-02, β=1.07E-01 and ε_2=1.09E-02);
- MOFS: case a (ρ=9.05E-01, μ=5.45E-02, β=9.00E-02 and ε_2=9.14E-03) and case b (ρ=1.09, μ=5.32E-02, β=1.09E-01 and ε_2=1.04E-02).

As observed in Fig. 3.5, where the best value considering the amplitude minimization is considered, the bandwidth is smaller than the one observed in Fig. 3.6. On the other hand, where the best value considering the bandwidth maximization is taken into account, this behavior is reversed, thus emphasizing the characteristic conflict of the Pareto's Curve.

As mentioned in Chapter 2, it becomes evident that all parameters analyzed contribute to the appearance of instabilities in the system. It can be seen instabilities regions ranging up to near Ω=0.926 until Ω=0.928, as observed in Fig. 3.5. In addition, in both figures is very important to observe the attenuation of the amplitude of vibration of nDVA in comparison with the DVA.

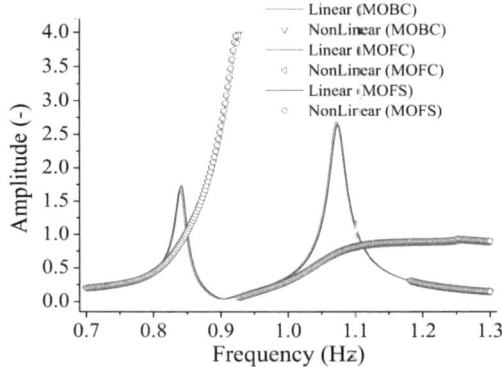

Figure 3.5 – Simulation of the nDVA considering the amplitude minimization by using the BiOM.

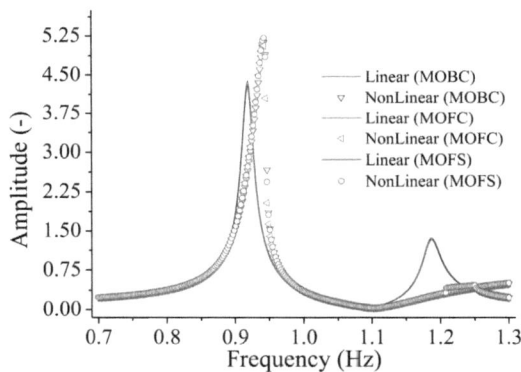

Figure 3.6 – Simulation of the nDVA considering the bandwidth maximization by using the BiOM.

3.6 SENSITIVITY ANALYSIS OF BiOM PARAMETERS

This section is dedicated to the sensitivity analysis of selected BiOM parameters in the Pareto's Curve. For design purposes, the parameters used by BiOM: MOBC - number of scout bees (5), number of bees recruited for the best e sites (5), number of bees recruited for the other selected sites (3), number of sites selected for neighborhood search (2), number of top-rated (elite) sites among m selected sites (5), neighborhood search (ngh) [10^{-3} 10^{-4} 10^{-6}] and generation number (100); MOFC - number of fireflies (20), maximum attractiveness value (0.9), absorption coefficient [0.7 0.9 1.0] and generation number (100); MOFS - number of fishes (20), weighted parameter value (1), control fish displacements [10^{-1} 10^{-2} 10^{-3}] and generation number (100). For these parameters, the number of objective function evaluations is 20+20×100. In this case, the neighborhood search (MOBC), the absorption coefficient (MOFC) and the control fish displacements (MOFS) were the parameters chosen for the sensitivity analysis. All other characteristics considered during the optimization process were previously defined in section 3.4.

Table 3.4 presents selected points ("best" value in terms of the maximization of the bandwidth and "best" value in terms of the minimization of the amplitude) of the Pareto's Curve obtained by the BiOM, considering different parameters. In this table it is possible to observe that the parameters analyzed had small influence on the shape of the Pareto's Curve.

Table 3.4 – Selected points (Extremes) of the Pareto's Curve by using the BIOM, considering different parameters.

		ρ	μ	β	ε_2	Amplitude	Bandwidth
MOBC	10^{-3}	9.03E-01	5.45E-02	9.00E-02	9.41E-03	4.67	4.02E-02
		1.09	5.29E-02	1.07E-01	1.09E-02	5.83	1.57E-01
	10^{-4}	9.01E-01	5.41E-02	9.00E-02	9.25E-03	4.65	4.02E-02
		1.09	5.28E-02	1.05E-01	1.09E-02	5.72	1.57E-01
	10^{-6}	9.05E-01	5.40E-02	9.00E-02	9.00E-03	4.65	4.02E-02
		1.09	5.31E-02	1.09E-01	1.07E-02	5.87	1.57E-01
MOFC	0.7	9.00E-01	5.41E-02	9.00E-02	9.32E-03	4.69	3.62E-02
		1.09	5.41E-02	1.07E-01	1.09E-02	5.86	1.57E-01
	0.9	9.15E-01	5.47E-02	9.00E-02	9.06E-03	4.66	5.23E-02
		1.09	4.93E-02	1.08E-01	1.07E-02	5.91	1.57E-01
	1.0	1.09	5.34E-02	1.08E-01	1.07E-02	5.88	1.57E-01
		9.06E-01	5.41E-02	9.00E-02	9.08E-03	4.65	4.42E-02
MOFS	10^{-1}	9.05E-01	5.45E-02	9.00E-02	9.14E-03	4.67	4.42E-02
		1.09	5.32E-02	1.09E-01	1.04E-02	5.91	1.57E-01
	10^{-2}	9.14E-01	5.46E-02	9.00E-02	9.08E-03	4.67	5.27E-02
		1.09	4.92E-02	1.09E-01	1.09E-02	5.91	1.57E-01
	10^{-3}	9.02E-01	5.40E-02	9.00E-02	9.23E-03	4.66	4.01E-02
		1.09	5.24E-02	1.03E-01	1.08E-02	5.71	1.57E-01

The numerical applications showed that the sensitivities of dynamic responses convey valuable information about the influence of the design parameters on the dynamic behavior of the nonlinear structure, being also a useful tool for design and analysis of modified systems and structural optimization. The choice of the design variables was based on previous knowledge regarding their sensitivities with respect to the amplitude peak and suppression bandwidth. It is worth mentioning that these parameters are directly associated with the effectiveness of the nDVA.

Finally, as demonstrated by the results, the nonlinearity factor is an important parameter to be investigated during the design procedure of nonlinear dynamic vibration absorbers, due to its contribution to the reduction of the vibration level. This point motivates an important procedure regarding the presented methodology, namely to obtain the optimal spring nonlinear coefficient that guarantees the best stable solution for a given system.

3.7 PRELIMINARY CONCLUSIONS

In this chapter, the numerical optimization of a two degree-of-freedom nonlinear damped system composed of a primary mass attached to the ground by a linear spring and a secondary mass attached to the primary system by a nonlinear spring was studied. The system design was oriented so that both the maximization of the attenuation frequency bandwidth and the minimization of the amplitude response were taken into account simultaneously. For this aim, three different algorithms, namely, BCA (MOBC - Multi-objective Optimization Bee Colony), Firefly Colony (MOFC - Multi-objective Optimization Firefly Colony) and Fish Swarm (MOFS - Multi-objective Optimization Fish Swarm) were tested. Each evolutionary approach is associated with the non-dominated sorting and crowding distance operators to obtain the optimal design of the system. As observed in Tabs. 3.2 to 3.4 the results show that the methodology used represents an interesting alternative to the treatment of the formulated optimization problem.

CHAPTER 4:
INVERSE PROBLEM – APPLICATION TO A NONLINEAR DYNAMIC VIBRATION ABSORBER USING BIO-INSPIRED OPTIMIZATION METHODS

4.1 INTRODUCTION

Inverse Problems (IP) is a research area dealing with the inversion of models or experimental data. Itarises from the necessity of determining parameters of theoretical models in such a way that theycan be used to simulate the behavior of the system for different operating conditions. In the most useful applications, theoretical models are fitted to experimental results. Inverse problems arise frequently in various engineering design problems, such as: process control, fermentation process, drying, medical imaging, structural engineering, aerospace engineering, geophysics, computer vision, astronomy, nondestructive testing, and many others. Consequently, an IP is a mathematical framework that is used to obtain information about a physical object or system from observed measurements. The solution to this problem is useful since it generally provides information about physical parameters that we cannot observe directly (Tarantola, 2005). The estimation procedure consists in obtaining the model parameters through the minimization of the difference between calculated and experimental values.

Figure 4.1 presents the flowchart encompassing Direct and Inverse Problems.

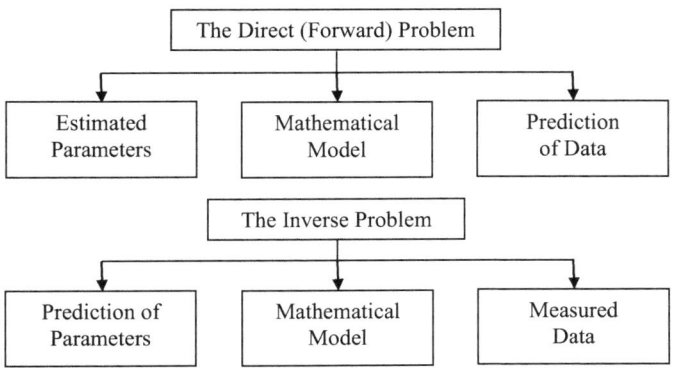

Figure 4.1 – Flowchart: Direct and Inverse Problems.

According to Tarantola (2005), the scientific procedure for the study of a physical system can be (rather arbitrarily) divided into the following three steps:

- Parameterization: discovery of a minimal set of model parameters whose values completely characterize the system studied;
- Direct modeling: discovery of the physical laws allowing to make predictions on the results of measurements on some observable parameters;
- Inverse modeling: use of the actual results of some measurements of the observable parameters to infer the actual values of the model parameters.

Traditionally, IP have been treated by two different approaches: the deterministic approach that makes use of the Variational Calculus, and the non-deterministic one that it is based on the process of natural selection, i.e., in the genetics of the populations or, alternatively, in purely structural methodologies. The use of the non-deterministic approach is getting increasing attention in the last decade, mainly because they do not use derivatives and they can be easily implemented in the computer. Besides, non-deterministic approaches are in general more robust than the deterministic ones when dealing with inverse problems.

In the DVA context, the literature brings valuable information. The expressions of matrix construction by using the singular value decomposition are applied to the physical parameter identification of the dynamic model (Li et al., 2002). Gungor et al. (2008) studied a large flexible beam with a tip mass-pendulum system to be effective vibration-absorbing device for large flexible structures with tip appendages. The technique of digital image processing was used to demonstrate how unique experimental results can be obtained by the use of high performance data acquisition and high speed cameras. Experimental results obtained from image processing were used as input to the inverse problem, to determine the pendulum mass, length and viscous damping, through the Levenberg-Marquardt Algorithm. Zhu et al. (2009) studied the inverse problem of assigning receptances to a dynamic system by using one or more simple mass-spring absorbers. The absorber parameters were determined by using selected receptances from the original system.

4.2 FORMULATION AND SOLUTION OF AN INVERSE PROBLEM

In this chapter we are interested on the determination of the following properties of a nonlinear dynamic vibration absorber: ρ, μ, β and ε_2. It should be mentioned that the choice of the most sensitive design variables was based on previous knowledge of their sensitivities (see Chapter 2).

Mathematically, the inverse problem is formulated as:

$$OF \equiv \min_{\rho,\mu,\beta,\varepsilon_2} \sum_{i=1}^{N} \left(A_i^{\text{exp}} - A_i^{\text{sim}} \right)^2 \tag{4.1}$$

where A^{exp} and A^{sim} are the experimental and simulated displacement amplitudes, respectively, and N represents the total number of experimental data. As the number of measured data, N, is usually much larger than the number of parameters to be estimated, the inverse problem is formulated as a finite dimensional optimization problem in which we aim at minimizing the Objective Function (OF).

As real experimental data were not available, we generated synthetic experimental data, as follows:

$$A_i^{\text{exp}} = A_i^{\text{sim}} + \sigma r \tag{4.2}$$

Where r is a random number from a Gaussian distribution with zero mean and unitary standard deviation, and σ stands for the standard deviation of measurement errors.

In order to evaluate the performance of the three optimization techniques proposed above (BCA, FCA, and FSA), the following parameters were used in the algorithms:

- BCA parameters: number of scout bees (10), number of bees recruited for the best e sites (5), number of bees recruited for the other selected sites (5), number of sites selected for neighborhood search (5), number of top-rated (elite) sites among m selected sites (5), neighborhood search (ngh) (10^{-3}), and generation number (100);
- FCA parameters: number of fireflies (15), maximum attractiveness value (0.9), absorption coefficient (0.7) and generation number (100);
- FSA parameters: number of fish (15), weighted parameter value (1), control fish displacements (10^{-1}), and generation number (100).

In order to examine the quality of the solution, the Genetic Algorithm (GA) has been performed for comparison purposes. The parameters used by GA are the following: population size (15), crossover rate(0.8), mutation rate (0.01), and generation number (100).

The stopping criterion used was the maximum number of iterations. Each case study was computed 20 times before calculating the average values. It is worth mentioning that 1515 objective function evaluations for each algorithm are necessary. For the amplitude, 300 experimental points were considered.

In order to examine the accuracy of the inverse problem methodology used for the estimation of the parameters (ρ, μ, β and ε_2),noise was taken into account in the tests (σ=0.02, i.e., corresponding to 7% error; or without noise, σ=0).

4.2.1 Test Case # 1 (ρ=1, μ=0.05, β=0.1 and ε_2=0.01)

Figure 4.2 presents a simulation (with and without noise) of the forcing frequency versus the displacement amplitude both for a linear and nonlinear dynamic vibration absorber considering the following parameters: ρ=1, μ=0.05, β=0.1 and ε_2=0.01.

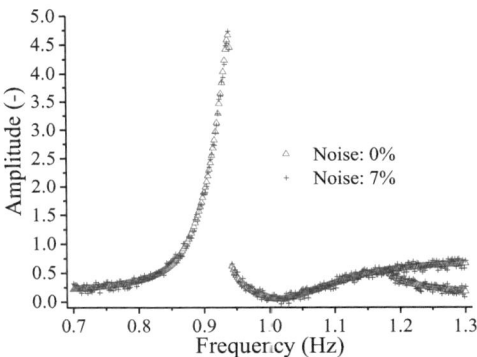

Figure 4.2 – Forcing frequency versus displacement amplitude considering the linear and nonlinear cases (ρ=1, μ=0.05, β=0.1 and ε_2=0.01). Two noise conditions evaluated: σ=0.02 and σ=0.

For the simulations, the following ranges are considered for the design parameters: $0.9 \leq \rho \leq 1.2$, $0.04 \leq \mu \leq 0.06$, $0.09 \leq \beta \leq 1.2$, and $0.009 \leq \varepsilon_2 \leq 0.012$. In Table 6.1 the best results obtained for the design of the nonlinear vibration absorber using BiOM considering noise (σ=0.02) and without noise (σ=0) are presented. This table shows that the BCA, FCA and FSA led to satisfactory estimates for the unknown parameters as compared with the GA, considering noiseless data (σ=0, see Table 4.1),as given by the values obtained for the objective function. However, when noise is added, the BCA, FCA and FSA leads to gcod estimates.

Table 4.1. Results obtained considering the BiOM and GA for test case # 1.

Method	Noise	ρ	μ	β	ε_2	OF
		1	0.05	0.1	0.01	-
BCA	0%	0.99988*	0.05000	0.09999	0.01010	2.012E-05
		(1.2E-04)**	(5.2E-04)	(3.4E-05)	(2.1E-04)	(2.8E-06)
	7%	0.99964	0.04992	0.10051	0.01002	0.92963
		(3.5E-03)	(7.7E-03)	(7.1E-03)	(1.1E-02)	(1.6E-02)
FCA	0%	0.99989	0.05001	0.09999	0.01000	2.112E-05
		(1.9E-05)	(3.2E-05)	(1.2E-05)	(4.1E-05)	(1.2E-06)
	7%	0.99961	0.04999	0.10051	0.01003	0.92964
		(5.6E-04)	(6.7E-03)	(5.6E-03)	(4.4E-02)	(7.5E-02)
FSA	0%	0.99999	0.05000	0.09999	0.01000	1.029E-05
		(3.8E-05)	(8.2E-05)	(1.6E-05)	(1.4E-05)	(1.2E-06)
	7%	0.99974	0.04996	0.10054	0.00997	0.92966
		(6.7E-03)	(6.6E-02)	(8.9E-03)	(3.7E-02)	(1.8E-03)
GA	0%	0.99989	0.04999	0.09999	0.01000	2.444E-05
		(1.1E-05)	(2.1E-05)	(1.1E-05)	(1.6E-05)	(1.2E-06)
	7%	0.99975	0.04995	0.10033	0.00996	0.92959
		(2.7E-02)	(9.9E-03)	(1.9E-03)	(9.9E-02)	(5.8E-02)

*Average value and **Standard deviation.

4.2.2 Test Case # 2 (ρ=1.2, μ=0.03, β=0.08 and ε_2=0.02)

For the second simulation, Figure 4.3 presents the results for the cases with and without noise of the forcing frequency versus the displacement amplitude for the linear and nonlinear dynamic vibration absorbers. The following parameters are considered: ρ=1.2, μ=0.03, β=0.08 and ε_2=0.02.

The following ranges are defined for the design parameters: 1.1 $\leq\rho\leq$ 1.3, 0.02 $\leq\mu\leq$ 0.05, 0.07 $\leq\beta\leq$ 0.09, and 0.01 $\leq\varepsilon_2\leq$ 0.03. The best results obtained for the design of the nonlinear vibration absorber using BiOMfor the cases with noise (σ=0.02) and without noise (σ=0) are presented in Table 4.2. In this table, for the noiseless data, it is possible to observe that the BCA, FCA and FSA were able to satisfactorily estimate the unknown parameters as compared with the results from GA (as demonstrated by the values obtained for the objective function). However, when noise is added, theBCA, FCA and FSA are able to obtain good estimates.

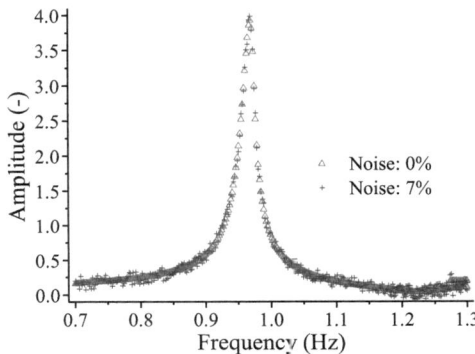

Figure 4.3 – Forcing frequency versus displacement amplitude considering the linear and nonlinear cases (ρ=1.2, μ=0.03, β=0.08 and ε_2=0.02). Two noise conditions evaluated:σ=0.02 andσ=0.

Table 4.2. Results obtained considering the BiOM and GA for test case # 2.

Method	Noise	ρ	μ	β	ε_2	OF
		1.2	0.03	0.08	0.02	-
BCA	0%	1.19999*	0.02999	0.07999	0.01999	3.099E-05
		(2.3E-04)**	(1.1E-05)	(7.1E-05)	(7.1E-05)	(1.8E-06)
	7%	1.19930	0.02994	0.07996	0.01919	1.02546
		(1.2E-03)	(7.4E-03)	(2.8E-02)	(1.8E-03)	(7.6E-02)
FCA	0%	1.19999	0.03010	0.07999	0.01999	4.095E-05
		(1.1E-05)	(2.7E-05)	(3.2E-05)	(2.4E-05)	(1.4E-06)
	7%	1.19942	0.02990	0.07993	0.01893	1.02678
		(3.7E-02)	(5.6E-03)	(7.9E-03)	(2.6E-03)	(8.7E-02)
FSA	0%	1.19999	0.02999	0.07999	0.02010	4.110E-05
		(1.6E-05)	(1.2E-05)	(5.6E-05)	(5.6E-05)	(2.9E-06)
	7%	1.19933	0.02990	0.07995	0.01917	1.02677
		(2.4E-02)	(2.7E-03)	(2.9E-03)	(1.7E-03)	(4.5E-02)
GA	0%	1.20001	0.03010	0.08010	0.01999	4.100E-05
		(7.7E-06)	(1.9E-05)	(1.8E-05)	(2.2E-05)	(2.3E-06)
	7%	1.19929	0.02974	0.07987	0.01922	1.02500
		(1.3E-02)	(4.7E-03)	(9.8E-03)	(3.4E-02)	(4.8E-02)

*Average value and **Standard deviation.

43

4.2.3 Test Case # 3 (ρ=0.8, μ=0.08, β=0.12 and ε_2=0.009)

In the last case, Figure 4.4 presents the results for the cases with and without noise regarding the forcing frequency versus the displacement amplitude for the linear and nonlinear dynamic vibration absorbers. The following parameters are consideredρ=0.8, μ=0.08, β=0.12 and ε_2=0.009.

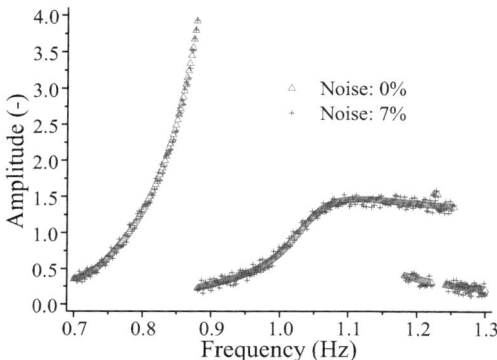

Figure 4.4 – Forcing frequency versus displacement amplitude considering the linear and nonlinear cases (ρ=0.8, μ=0.08, β=0.12 and ε_2=0.009). Two noise conditions evaluated: σ=0.02 and σ=0.

The following ranges are defined for the design parameters: $0.9 \leq \rho \leq 1.1$, $0.07 \leq \mu \leq 0.09$, $0.11 \leq \beta \leq 0.13$, and $0.007 \leq \varepsilon_2 \leq 0.010$. In Table 4.3 the best results obtained for the design of the nonlinear vibration absorber using BiOM considering noisy (σ=0.02) and without noise (σ=0) are presented. As before, for the noiseless data, it ispossible to observe that the BCA, FCA and FSA were able to satisfactorily estimate the unknown parameters as compared with the results from GA (as demonstrated by the values obtained for the objective function). However, when noise is added, the BCA, FCA and FSA are able to obtain good estimates.

Table 4.3. Results obtained considering the BiOM and GA for test case 3.

Method	Noise	ρ	μ	β	ε_2	OF
		0.8	0.08	0.12	0.009	-
BCA	0%	0.79999[*]	0.08001	0.11999	0.00899	2.191E-05
		(1.3E-05)[**]	(2.3E-05)	(4.4E-05)	(2.2E-05)	(3.7E-06)
	7%	0.80196	0.07993	0.12139	0.00894	1.05012
		(3.3E-02)	(5.6E-02)	(7.1E-03)	(3.7E-03)	(2.3E-02)
FCA	0%	0.79999	0.08000	0.12001	0.00899	1.931E-05
		(4.4E-05)	(1.1E-05)	(2.1E-05)	(5.1E-05)	(1.3E-06)
	7%	0.80138	0.08006	0.12091	0.00896	1.05099
		(1.7E-02)	(3.4E-03)	(2.3E-03)	(3.3E-03)	(7.7E-02)
FSA	0%	0.79999	0.08000	0.12002	0.00900	3.098E-05
		(2.2E-05)	(3.1E-05)	(5.2E-05)	(5.1E-05)	(4.4E-06)
	7%	0.80195	0.07995	0.12116	0.00894	1.05032
		(1.1E-03)	(2.1E-02)	(6.7E-03)	(3.8E-03)	(6.3E-02)
GA	0%	0.79999	0.08000	0.12001	0.00900	2.991E-05
		(2.2E-05)	(1.1E-05)	(7.2E-05)	(6.7E-05)	(2.4E-06)
	7%	0.80342	0.07996	0.12216	0.00891	1.05414
		(1.3E-03)	(5.6E-02)	(2.2E-03)	(3.4E-03)	(1.2E-02)

*Average value and **Standard deviation.

4.3 PRELIMINARY CONCLUSIONS

In this chapter, the Bees Colony Algorithm, the Firefly Colony Algorithm, and the Fish Swarm Algorithm were proposed as alternative techniques to solve an inverse problem associated with the design of a dynamic vibration absorber. For illustration purposes, the algorithms were applied to the design of a nonlinear dynamic vibration absorber considering different configurations for the design variables. The system nonlinearity was introduced in the springs that connect the primary mass to the ground and the absorber to the primary mass, respectively.

As observed in Tables 4.1 to 4.3, the algorithms led to satisfactory results in terms of the values of the objective function as compared with the results from the GA strategy. However, the results obtained by the algorithms need yet to be better analyzed, so that final conclusions can be drawn. For example, new mechanisms for diversity exploration should be included. In addition, when noise is considered, the value of the objective function increases, as observed in Tables 4.1 to 4.3.

The choice of the design variables is based on previous knowledge regarding their sensitivities with respect to the displacement amplitude. It is worth mentioning that these parameters are directly associated with the effectiveness of then DVA.

In terms of the system resolution, the equations of motion of the nonlinear two-degree-of-freedom system were integrated numerically by using the so-called average method that provides an approximate solution to nonlinear dynamic problems.

The nonlinear algebraic equations were solved numerically enabling to determine the roots of the nonlinear algebraic equations. It is worth mentioning that the nonlinearity factor is an important parameter to be investigated during the design procedure of nDVAs, due to its contribution to the reduction of the vibration level. However, care must be taken in the context of high nonlinearity because of the instabilities introduced in the system. This point motivates an important goal regarding the proposed methodology, namely, to obtain the optimal spring nonlinear coefficient that guarantees the best stable solution for a given system.

CHAPTER 5:
ROBUST DESIGN OF A NONLINEAR DYNAMIC VIBRATION ABSORBER USING BIO-INSPIRED OPTIMIZATION METHODS

5. 1 INTRODUCTION

Real-world optimization problems contain uncertain data, inherently stochastic/random or it can be due to errors (Prékopa, 1995). According to Gorissen et al. (2015), the reasons for data errors are associated to measurement/estimation errors that come from the lack of knowledge of the parameters of the mathematical model or could result from implementation errors that come from the physical impossibility to exactly implement a computed solution in a real-life setting.

Traditionally, in the specialized literature two approaches to deal with data uncertainties in optimization can be found (Robust and Stochastic Optimization). Stochastic Optimization is based on the knowledge of the probability distribution of uncertain data (Prékopa, 1995). On the other hand, Robust Optimization is defined as an approach that produces a solution that is not sensitive to small changes in the design variables (Taguchi, 1984; Bertsimas et. al., 2011). It important to mention that a robust solution cannot coincide with the nominal one (solution without robustness, as observed in Fig. 5.1). In this context, robustness characterizes an important design characteristic for the solution, under certain conditions, when exposed to given conditions of uncertainty.

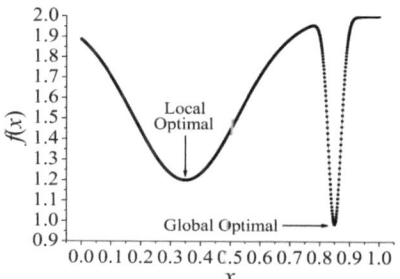

Figure 5.1 – Robust and sensitive solutions considering perturbations (Deb and Gupta, 2006).

In the classical deterministic optimization process, the uncertainties of the design variables are not taken into account (Borges et. al., 2010). Regarding the robustness evaluation of the optimal solutions, the problem posed is not to find only the optimum but the robust optimum by taking into account the uncertainties found along the optimization process. In nonlinear dynamics, uncertainties result from several sources, including the errors associated to the modeling of nonlinear physical phenomena, the characteristics of the materials (such as the Young's modulus, mass density, Poisson ratio, etc.), the geometry characteristics, the tolerances in the manufacturing processes (thickness, stiffness, etc). Classically, the robustness of the optimal solution is evaluated starting from the deterministic optimization process, supposing that the deterministic design space contains the robust solutions obtained through the inclusion of stochastic criteria.

In the literature, various works deal with robust optimization as applied to a number of applications. Lee and Park (1996a) proposed a similar procedure by using the Taguchi method, in which the robustness of the optimal solutions is evaluated only at the end of the optimization process. Lee and Park (1996b) proposed an optimization methodology through which an additional objective function is defined, having the same mean and standard deviation as the original objective function, by using a weighting objective method. The main disadvantage of the proposed methodology is that not all the Pareto optimal solutions can be found, unless the problem is convex. Ben-Tal and Nemirovski (2002) analyzed the performance of an antenna array when subjected to slight perturbations in the nominal data. The array was optimized to attenuate side lobes, but small implementation errors in the design variables cause the radiation pattern of the antenna to worsen dramatically in the region of interest. Borges et al. (2010) determined the robust optimal design of nonlinear dynamic vibration absorbers through the minimization of the vibration amplitude and the maximization of the suppression bandwidth. The nonlinearities were introduced both in the springs that connect the primary mass to the ground, and the absorber to the primary mass. Each robustness function is associated to an original objective function and is defined as being inversely proportional to the dispersion. Uncertainties on the design variables that characterize the nonlinear dynamic system are introduced directly through a parametric approach, by performing a LHC sampling method. Fang et al. (2015) studied the robustness and the performance of the fatigue life involving the material uncertainties and the structural design of a truck cab by using a multi-objective particle swarm optimization. These authors proposed a more general dual surrogate method, in which different dual response surface method, namely dual polynomial response surface, dual Kriging and dual radial basis function models were employed to fit the mean value and standard deviation of the fatigue life. Lan et al. (2015) presented a robust optimization model for designing a water supply system considering the risk of facility failure, and showed that it is equivalent to a mixed-integer linear program. The optimization problem formulated was then solved by using a Benders decomposition method.

In the Chapter 3, the nominal optimal design of a nDVA constituted of a primary mass attached to the ground by a linear spring and a secondary mass attached to the primary system by a nonlinear spring (see Fig. 2.1 in Chapter 2) by the using BiOM was considered. For this purpose, the maximization of the attenuation bandwidth (mono-objective problem) and the maximization of the attenuation bandwidth and the minimization of the amplitude (multi-objective problem) of the two degree-of-freedom nonlinear damped system are considered. For design purposes, it was considered that the mathematical model, the design variables and the parameters are sufficiently reliable, i.e., there are no errors of modeling and estimation. However, systems to be optimized are generally sensitive to small changes in the design variables leading to significant changes in the vector of objective functions. In this case, it is necessary to consider an approach for modeling optimization problems under uncertainty, where the modeler aims at finding decisions that are optimal for the worst-case realization of the uncertainties within a given set of values.

In this chapter, a different methodology to evaluate the robustness of the optimal solutions applied to nDVA design is presented. This methodology is based on the Mean Effective Concept as proposed by Deb and Gupta (2006).

5.2 MEAN EFFECTIVE CONCEPT

Traditionally, the introduction of robustness requires the addition of new restrictions and/or new objectives (relations between the mean and the standard deviation of the objective functions vector) and probability distribution functions for the design variables and/or objectives (Soize, 2001; 2005; Cataldo et. al., 2007; 2008; 2009; Sampaio and Soize, 2007; Ritto et. al., 2008).

As an alternative to these classical formulations, Deb and Gupta (2006) extended the Mean Effective Concept (MEC), originally proposed for mono-objective problems, to the multi-objective context. For this aim, the problem is rewritten as a mean vector of original objectives, i.e., no additional restriction is inserted into the original problem. Recently, Souza et al. (2015) applied this concept to solve optimal control problems by using the Multi-objective Optimization Differential Evolution. This methodology was used to solve singular optimal control problems with different levels of complexity, where the original continuous control trajectory is approximated by linear functions on time intervals.

Definition 5.1 - Mean Effective Concept (Deb and Gupta, 2006) – For the minimization of an objective function vector ($f(x)$), a solution x^* is called a robust solution if it is the global minimum of the mean effective function $f^{eff}(x)$, defined with respect to a δ-neighborhood as follows:

$$\min \left(\frac{1}{|\Upsilon_\delta(x)|} \int\limits_{y \in \Upsilon_\delta(x)} f_1 dy, \ \frac{1}{|\Upsilon_\delta(x)|} \int\limits_{y \in \Upsilon_\delta(x)} f_2 dy, \ \dots, \ \frac{1}{|\Upsilon_\delta(x)|} \int\limits_{y \in \Upsilon_\delta(x)} f_M dy \right) \quad (5.1)$$

where Y_δ is the δ-neighborhood of the solution x and $|Y_\delta|$ is the hyper volume of the neighborhood.

Basically, a finite set of solutions H can be generated randomly using, for example, the Latin Hypercube for the evaluation of the integral expressed by Eq. (5.1). Consequently, defining the δ-neighborhood with respect to the design variable vector, N_H solutions are generated employing the Latin Hypercube, with the integrals calculated numerically. This operation increases the computational cost due to number of integral calculations, necessary to evaluate the objective function (Deb and Gupta, 2006). To evaluate the uncertainties on the design variables, one considers the parametric approach by assuming that all design variables exhibit normal distributions with zero mean and unit variance.

5.3 ROBUST MULTI-OBJECTIVE OPTIMIZATION OF THE nDVA

The methodology proposed in this chapter consists of the following steps, as illustrated in Fig. 5.2.

- The first step consists of BiOM parameters definition. Next, a population of N candidates is generated randomly. All dominated solutions are removed from the population through the operator Fast Non-Dominated Sorting. In this way, the population is sorted into non-dominated fronts F_j (sets of vectors that are non-dominated with respect to the others). This procedure is repeated until each vector belongs to a front. An offspring is generated from the three parents (this process continues until N children are generated). The new population is then classified according to the dominance criterion. If the number of individuals of this population is larger than a number defined by the user, it is truncated according to the criterion named the Crowding Distance (Deb, 2001). More details about the MOBC (Multi-objective Optimization Bee Colony), MOFC (Multi-objective Optimization Firefly Colony) and MOFS (Multi-objective Optimization Fish Swarm) algorithms can be found in Section 3.3.

- In addition, for each individual of the population, N_H samples are generated randomly considering the δ-neighborhood of the current solution by using the Latin Hypercube for the evaluation of the integral expressed by Eq. (5.1). For each iteration, a new population with $N \times N_H$ candidates should be evaluated according to the objective function. For each candidate, the mean of these N_H evaluations is computed and this value is considered as the response to this N_H perturbations.

- This procedure is repeated until the maximum number of generations is reached.

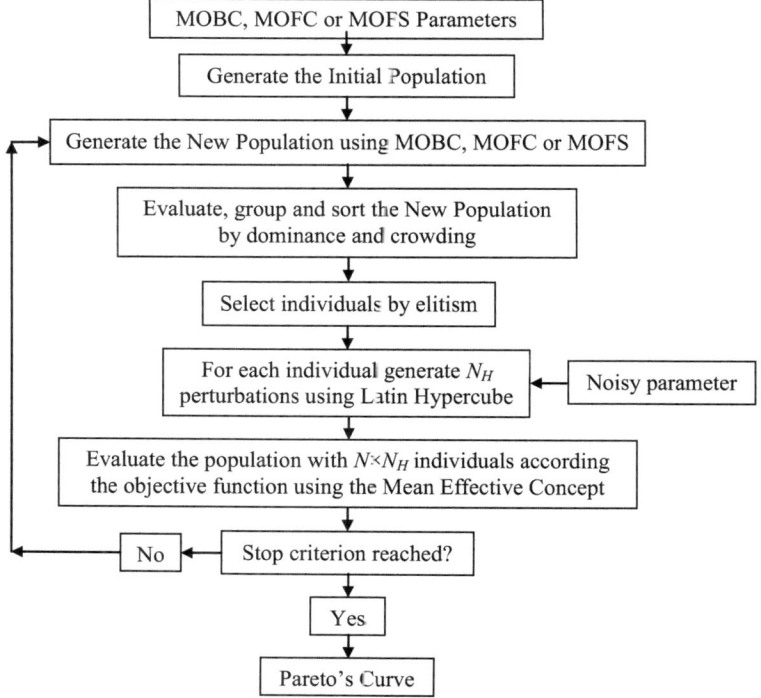

Figure 5.2 – Flowchart Robust BIOM.

For design purposes, the following steps are established:

- The optimization problem considered in this chapter is composed by the two following objective functions: *i*) minimization of the amplitude of the dynamic response of the nonlinear damped system corresponding to the first mode; and *ii*) the maximization of the suppression bandwidth;

- Design variables (normalized structural parameters): $0.8 \leq \rho \leq 1.2$, $0.045 \leq \mu \leq 0.08$, $0.08 \leq \beta \leq 0.12$ and $0.009 \leq \varepsilon_2 \leq 0.02$.

- The parameters used by MOFA: number of fireflies (25), maximum attractiveness value (0.9), absorption coefficient (0.7) and generation number (100). It should be noted that as all the BiOM algorithms presented similar results and the robust solution requires even more objective function evaluations in comparison with the nominal problem, the MOFA algorithm is chosen to obtain the robust design;

- Stopping criterion: maximum number of generations (100);

- Considering the parameters presented, to solve the nominal case by using the MOFC, $25+25\times100$ objective function evaluations are necessary. To solve the robust case by the MOFC, $25+25\times N_H\times100$ objective function evaluations are necessary (N_H=10 points considered in the Latin Hypercube);

- In this test case the following values for the parameter δ were considered: [1% 2.5% 5%];

- The computation consists in obtaining the driving point dynamic responses $H(\omega, p)$ associated to the displacement x_1 as indicated on Fig. 2.1 (see Chapter 2) in the frequency band of interest, namely Ω=[0.7–1.3 Hz], comprising a total of 300 frequency points. For the numerical solution of Eq. (2.8) (see Chapter 2) the following initial conditions were adopted (u_1=5, u_2=5, v_1=2 and v_2=4) by using the Newton Method.

Figure 5.3 shows the MOFA solutions obtained by applying the proposed robust method in comparison with the nominal solutions presented in Chapter 3. In practice, these functions consist in minimizing the dispersion around each cost function. Through this figure, one can notice that the intervals of dispersion for the cost functions are as follows: from approximately 4.7 to 6 for the optimal solutions corresponding to the first cost function, and from approximately 0.025 to 0.165 for the optimal solutions corresponding to the second cost function.

As observed in Fig. 5.3, increasing the value of the parameter δ, the distance from the Robust Pareto's Curve (1%, 2.5%, 5%) with respect to the nominal Pareto's Curve is also increased. The distance between these two curves is more significant for amplitude values smaller than approximately 5.05.

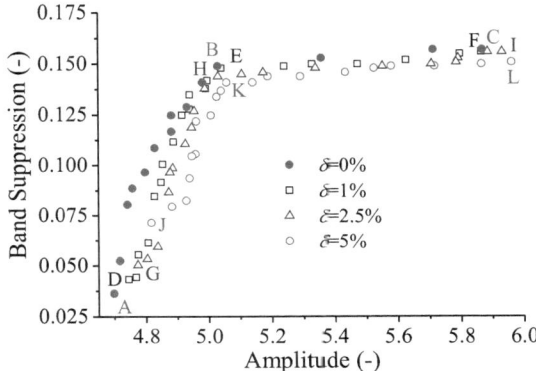

Figure 5.3 – Nominal and Robust Pareto's Curves by using the MOFC strategy.

Table 5.1 presents selected points from the nominal and robust Pareto's curves obtained by the MOFC.

Table 5.1 – Selected points from the nominal and robust Pareto's curves by using the MOFC.

δ		ρ	μ	β	ε_2	Amplitude	Bandwidth
	A	9.00E-01	1.07E-02	9.00E-02	9.32E-03	4.69	3.62E-02
0%	B	1.09	5.41E-02	9.16E-02	1.08E-02	5.02	1.48E-01
	C	1.09	5.41E-02	1.07E-01	1.09E-02	5.86	1.57E-01
	D	9.13E-01	5.10E-02	9.00E-02	9.03E-03	4.74	4.32E-02
1%	E	1.09	5.46E-02	9.12E-02	1.09E-02	5.04	1.47E-01
	F	1.09	5.36E-02	1.07E-01	1.06E-02	5.86	1.56E-01
	G	9.18E-01	5.47E-02	9.14E-02	9.66E-03	4.77	5.03E-02
2.5%	H	1.09	5.48E-02	9.12E-02	1.09E-02	5.03	1.43E-01
	I	1.09	5.37E-02	1.08E-01	1.07E-02	5.93	1.56E-01
	J	9.33E-01	5.34E-02	9.04E-02	9.07E-03	4.81	7.14E-02
5%	K	1.09	5.40E-02	9.01E-02	1.07E-02	5.05	1.40E-01
	L	1.09	5.22E-02	1.09E-01	1.07E-02	5.96	1.51E-01

Figure 5.4 presents the results from numerical simulation considering the points B, E, H and K (good compromise for amplitude minimization and bandwidth maximization) for different values of δ.

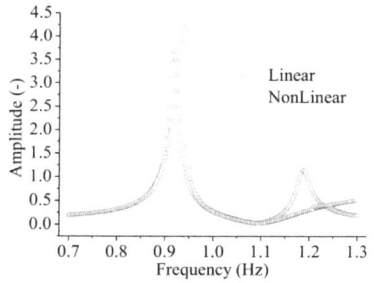

a) Simulation considering $\rho=1.09$, $\mu=5.41\text{E-}02$, $\beta=9.16\text{E-}02$ and $\varepsilon_2=1.08\text{E-}02$ (Nominal Optimal ($\delta=0$) – Point B).

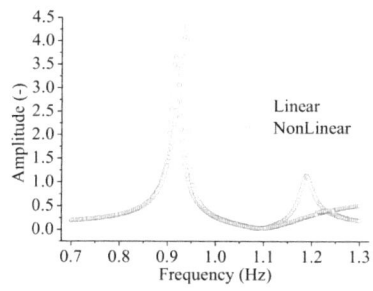

b) Simulation considering $\rho=1.09$, $\mu=5.46\text{E-}02$, $\beta=9.12\text{E-}02$ and $\varepsilon_2=1.09\text{E-}02$ (Robust Optimal ($\delta=1\%$) – Point E).

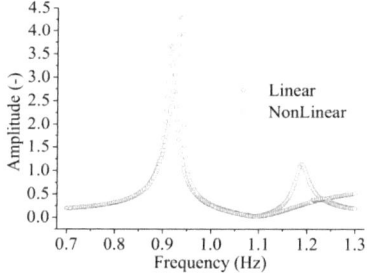

c) Simulation considering $\rho=1.09$, $\mu=5.48\text{E-}02$, $\beta=9.12\text{E-}02$ and $\varepsilon_2=1.09\text{E-}02$ (Robust Optimal ($\delta=2.5\%$) – Point H).

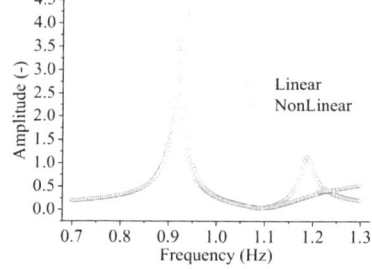

d) Simulation considering $\rho=1.09$, $\mu=5.40\text{E-}02$, $\beta=9.01\text{E-}02$ and $\varepsilon_2=1.07\text{E-}02$ (Robust Optimal ($\delta=5\%$) – Point K).

Figure 5.4 – Numerical simulations considering the points B, E, H and K.

As observed in Fig. 5.4, the simulation for the points B, E, H and K leads to similar curves. On the other hand, when points A, D, G and J (amplitude minimization) are considered, the simulation of the nominal solution is very different from those obtained by robust calculation, as presented in Fig. 5.5. However, these robust solutions are different with respect to the parameter δ considered, as shown in Fig. 5.6.

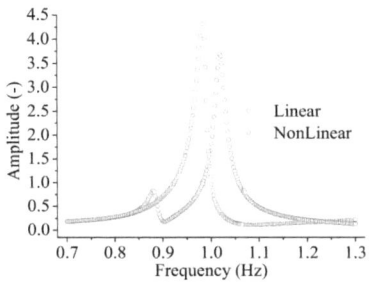

a) Simulation considering ρ=9.00E-01, μ=1.07E-02, β=9.00E-02 and ε_2=9.32E-02 (Nominal Optimal (δ=0) – Point A).

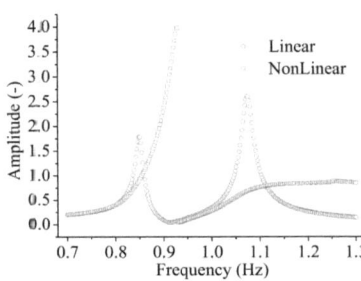

b) Simulation considering ρ=9.13E-01, μ=5.10E-02, β=9.00E-02 and ε_2=9.03E-03 (Robust Optimal (δ=1%) – Point D).

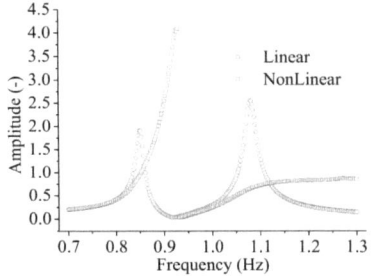

c) Simulation considering ρ=9.18E-01, μ=5.47E-02, β=9.14E-02 and ε_2=9.66E-03 (Robust Optimal (δ=2.5%) – Point G).

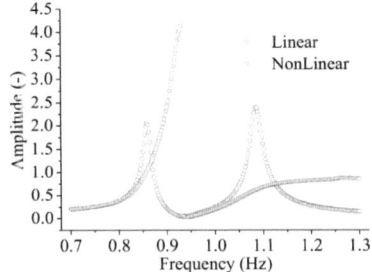

d) Simulation considering ρ=9.33E-01, μ=5.34E-02, β=9.04E-02 and ε_2=9.07E-03 (Robust Optimal (δ=5%) – Point J).

Figure 5.5 – Simulation considering the points A, D, G and J.

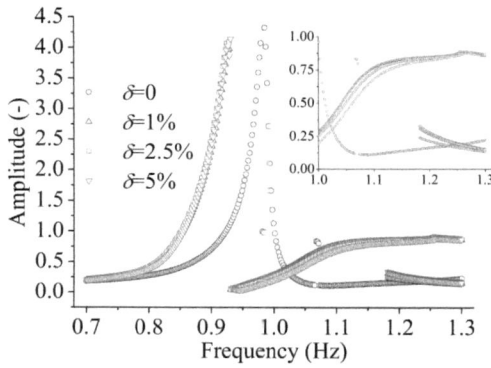

Figure 5.6 – Nominal and Robust simulations considering the points A, D, G and J.

5.4 PRELIMINARY CONCLUSIONS

In this chapter, the robust optimal design of nonlinear dynamic vibration absorbers was proposed and implemented by using the Multi-objective Optimization Firefly Algorithm, where the nonlinearities were introduced both in the springs that connect the primary mass to the ground, and the absorber to the primary mass. For this purpose, the Mean Effective Concept was considered to insert uncertainties in the design variables that characterize the nonlinear dynamic system. This is done directly through a parametric approach, by performing a Latin Hypercube sampling method. In general, it was shown that the mean effective concept can be easily incorporated into the Multi-objective Optimization Firefly Algorithm. However, the main disadvantage of this approach is the increase of the number of objective function evaluations necessary to calculate the integral that appears in the mean effective concept, being this approach independent from the optimization strategy considered.

As demonstrated by the results, δ is an important parameter to be investigated during the robust design procedure of nonlinear dynamic systems, due to its contribution to build the Pareto's Curve. As observed in the test case, the increase of δ parameter increases the distance between the robust and nominal Pareto's Curves. This point motivates an important procedure regarding the presented methodology, i.e., to obtain the optimal spring nonlinear coefficient that guarantees the best stable solution for a given system.

Finally, the proposed robust optimal design strategy demonstrates the importance of introducing uncertainties in the design variables to be evaluated during the optimization process in order to determine the optimal robust design that leads to the most effective nDVA for a given application.

CHAPTER 6:
CONCLUSIONS

This book brings a contribution about the optimal design of mechanical systems. For illustration purposes, a two degree-of-freedom nonlinear damped system composed of a primary mass attached to the ground by a linear spring and a secondary mass attached to the primary system by a nonlinear spring, thus forming a nonlinear dynamic vibration absorber, was used.

Chapter 2 presented the simulation and sensibility analysis of a nonlinear dynamic vibration absorber. The equations of motion of the nonlinear two d.o.f. system were numerically integrated by using the so-called average method that provides an approximate solution of nonlinear dynamic problems. The nonlinear algebraic equations obtained were solved numerically by using the Newton Method. It is worth mentioning that the study of the nonlinearity factor is very important due to its contribution to the reduction of the vibration level. However, care must be taken with high nonlinearities because of the instabilities introduced in the system.

Chapter 3 presents the design of a nDVA in which both the minimization of the amplitude and the maximization of the suppression bandwidth were taken into account. To solve the corresponding multi-objective optimization problem, three Bio-inspired Optimization Methods, namely the Bee Colony Algorithm, the Firefly Colony Algorithm and the Fish Swarm Algorithm were tested. Each evolutionary approach is associated with the non-dominated sorting and crowding distance operators to obtain the optimal design of the system. The results obtained show that the proposed methodology represents an interesting approach to the treatment of the optimization problem formulated as compared with those obtained by the Non-dominated Sorting Genetic Algorithm. It is important to observe that the present methodology eliminates the necessity of transforming the original multi-response optimization problem into a similar single-objective one, i.e., the original multi-objective optimization problem is directly solved.

Chapter 4 studied the solution of an inverse problem by using the following Bio-inspired optimization methods: Bees Colony Algorithm, Firefly Colony Algorithm and Fish Swarm Algorithm. As observed in Tables 4.1 to 4.3, the algorithms led to satisfactory results in terms of the values of the objective functions as compared with the GA strategy. Adding noise it is possible to observe that the BCA, FCA and FSA techniques were able to estimate the unknown parameters properly, as compared with the GA. This was demonstrated by the values obtained for the optimal objective function in the different cases.

Finally, Chapter 5 presented the robust optimal design of the nDVA by using the Multi-objective Optimization Firefly Algorithm. In this context, the Mean Effective Concept was used to insert uncertainties in the design variables that characterize the nonlinear dynamic system, by performing a Latin Hypercube sampling method. As demonstrated by the results, the increase of the δ parameter increases the distance from the robust Pareto's Curve with respect to the nominal Pareto's Curve. This point permits to obtain the value of the optimal spring nonlinear coefficient that guarantees the best stable solution for a given system. In addition, it should be mentioned that the mean effective concept can be easily incorporated into the Multi-objective Optimization Firefly Algorithm. However, the main disadvantage of this approach is the increase of the number of objective function evaluations necessary to evaluate the integral considered in the mean effective concept, independently from the chosen optimization strategy.

ACKNOWLEDGMENTS

The authors acknowledge the financial support provided by FAPEMIG (Fundação de Amparo à Pesquisa do Estado de Minas Gerais), CNPq (Conselho Nacional de Desenvolvimento Científico e Tecnológico) and FAPEG (Fundação de Amparo à Pesquisa do Estado de Goiás). Prof. Steffen acknowledges the financial support provided by FAPEMIG and CNPq (INCT-EIE, Instituto Nacional de Ciência e Tecnologia de Estruturas Inteligentes em Engenharia).

REFERENCES

Afshar, A., Haddad, O. B., Mariño, M. A., Adams, B. J. Honey-Bee Mating Optimization (HBMO) Algorithm for Optimal Reservoir Operation. *Journal of the Franklin Institute*, 344, 452–462, 2001.

Apostolopoulos T., Vlachos, A. Application of the Firefly Algorithm for Solving the Economic Emissions Load Dispatch Problem. *International Journal of Combinatorics*, 1–23, 2011.

Azeem, M. F., Saad, A. M. Modified Queen Bee Evolution Based Genetic Algorithm for Tuning of Scaling Factors of Fuzzy Knowledge Base Controller. IEEE INDICON 2004, Proceedings of the India Annual Conference, 299–303, 2004.

Ben-Tal, A., Nemirovski, A. Robust Optimization – Methodology and Applications, *Mathematical Programming,* 92, 453–480, 2002.

Bertsimas, D., Brown, D. B., Caramanis, C. Theory and Applications of Robust Optimization. *SIAM Review*, 53(3), 464–501, 2011.

Bonsel, J. H. Application of a Dynamic Vibration Absorber to a Piecewise Linear Beam System. Master's Thesis, Eindhoven University of Technology (TU/e) Department of Mechanical Engineering, 2003.

Borges, R. A., Lima, A. M. G., Steffen Jr., V. Robust Optimal Design of a Nonlinear Dynamic Vibration Absorber Combining Sensitivity Analysis. *Shock and Vibration*, 17 (4-5), 507–520, 2010.

Borges, R. A., Lobato, F. S., Steffen Jr., V. Application of Three Bioinspired Optimization Methods for the Design of a Nonlinear Mechanical System. *Mathematical Problems in Engineering*, 2, 1–12, 2013.

Cai, Y. Artificial Fish School Algorithm Applied in a Combinatorial Optimization Problem. *Intelligent Systems and Applications*, 1, 37–43, 2010.

Camazine, S., Deneubourg J., Franks N. R., Sneyd J., Theraula G., Bonabeau E. Self-Organization in Biological Systems, Princeton University Press, 2003.

Cataldo, E., Soize, C., Desceliers, C. and Sampaio, R. Uncertainties in Mechanical Models of Larynx and Vocal Tract for Voice Production, Proceedings of the XII DINAME, 2007.

Cataldo, E., Sampaio, R., Lucero, J., Soize, C. Modeling Random Uncertainties in Voice Production using a Parametric Approach. *Mechanics Research Communications*, 35(7), 454–459, 2008.

Cataldo, E., Soize, C., Sampaio, R., Desceliers, C. Probabilistic Modeling of a Nonlinear Dynamical System used for Producing Voice. *Computational Mechanics*, 43(2), 262-275, 2009.

Chang, H. S. Converging Marriage in Honey-Bees Optimization and Application to Stochastic Dynamic Programming. *Journal of Global Optimization*, 35, 423–441, 2006.

Cheung, Y. L., Wong, W. O., Cheng, L. Design Optimization of a Damped Hybrid Vibration Absorber. *Journal of Sound and Vibration*, 331, 750–766, 2012.

Cheung, Y. L., Wong, W. O., Cheng, L. Optimization of a Hybrid Vibration Absorber for Vibration Control of Structures under Random Force Excitation. *Journal of Sound and Vibration*, 332, 494–509, 2013.

Cunha Jr., S. S. Theoretical and Numeric Study of Dynamic Vibration Absorbers, M. Sc. Dissertation, Federal University of Uberlândia, Uberlândia, MG, Brazil, 1999.

Deb, K. Multi-Objective Optimization using Evolutionary Algorithms, John Wiley & Sons, Chichester, UK, ISBN 0-471-87339-X, 2001.

Deb, K., Agrawal, S., Pratab, A., Meyarivan, T. A Fast Elitist Non-Dominated Sorting Genetic Algorithm for Multi-Objective Optimization: NSGA-II. Kanpur - India, 2000.

Deb, K., Gupta, H. Introducing Robustness in Multi-Objective Optimization. *Evolutionary Computation*, 14, 463–494, 2006.

Edgeworth, F. Y. Mathematical Psychics, P. Keagan, London, England, 1881.

Eschenauer, J., Koski, J. and Osyczka, A. Multicriteria Design Optimization, Springer-Verlag, 1990.

Espíndola, J. J., Bavastri, C. A., Viscoelastic Neutralisers in Vibration Abatement: A Non-linear Optimization Approach. *Brazilian Journal of Mechanical Sciences*, 19(2), 154–163, 1997.

Espíndola, J. J., Bavastri, C. A., Oliveira-Lopes, E. M. Design of Optimum Systems of Viscoelastic Vibration Absorbers for a given Material based on the Fractional Calculus Model. *Journal of Vibration and Control*, 14(9-10), 1607–1630, 2008.

Espíndola, J. J., Pereira, P., Bavastri, C. A., Lopes, E. M., Design of Optimum System of Viscoelastic Vibration Absorbers with a Frobenius Norm Objective Function. *Journal of the Mathematical Problems in Engineering*, 11, Brazilian Society of Mechanical Sciences and Engineering, 31(3), 210–219, 2009.

Fang, J., Gao, Y., Sun, G., Xu, C., Li, Q. Multiobjective Robust Design Optimization of Fatigue Life for a Truck Cab. *Reliability Engineering and System Safety*, 135, 1–8, 2015.

Febbo, M., Machado, S. P. Nonlinear Dynamic Vibration Absorbers with a Saturation. *Journal of Sound and Vibration*, 332, 1465–1483, 2013.

Frahm, H. Device for Damping Vibrations of Bodies, US Patent 989, 958, 1911.

Gorissen, B. L., Yanıkoğlu, İ., den Hertog, D. A Practical Guide to Robust Optimization. *Omega*, 53, 124–137, 2015.

Gungor, F., Gumus, E., Ertas, A., Ekwaro-Osire, S., Nieto, E. Vibration Absorption of Tip Appendage Using Digital Image Processing Proceedings of the XIth International Congress and Exposition June 2-5, Orlando, Florida USA, 2008.

Hartog, J. P. D.. Mechanical Vibrations, McGraw-Hill Book Company, 1934.

Haug, E. J., Choi, K. K., Komkov, V. Design Sensitivity Analysis of Structural Systems, Academic Press, 1986.

Koronev, B. G., Reznikov, L. M. Dynamic Vibration Absorbers: Theory and Technical Applications, John Wiley & Sons, 1993.

Lan, F., Lin, W. H., Lansey, K. Scenario-based Robust Optimization of a Water Supply System under Risk of Facility Failure. *Environmental Modelling & Software*, 67, 160–172, 2015.

Lee, K. H., Park, G. J. Robust Design for Unconstrained Optimization Problems using the Taguchi Method. *AIAA Journal*, 34(5), 1059–1063, 1996a.

Lee, K. H., Park, G. J. Robust Optimization Considering Tolerance of Design Variables. *Journal of Computer and Structures*, 79(1), 77–86, 1996b.

Li, X. L., Shao, Z. J., Qian, J. X. An Optimizing Method based on Autonomous Animate: Fish Swarm Algorithm. *System Engineering Theory and Practice*, 22(11), 32–38, 2002.

Li, S., Zhang, F. Wang, B., Zhang, X. Proper Application of a Kind of Matrix Construction Method in Physical Parameter Identification of Dynamic Model. *Applied Mathematics and Mechanics*, 23(5), 606–613, 2002.

Li, X. L., Xue, Y. C., Lu, F., Tian, G. H. Parameter Estimation Method based on Artificial Fish School Algorithm. *Journal of Shan Dong University* (Engineering Science), 34 (3), 84–87, 2004.

Lima, A. M. G., Stoppa, M. H., Rade, D. A., Steffen Jr., V. Sensitivity Analysis of Viscoelastic Systems. *Shock and Vibration*, 13(4), 545–558, 2006.

Lima, A. M. G. Modelling and Robust Optimisation of Viscoelastic Damping in Mechanical Systems. Doctorate Thesis, University of Franche-Comté (UFC), Besançon in France, and Federal University of Uberlândia (UFU), Uberlândia in Brazil (in French), 2007.

Lobato, F. S., Sousa, J. A., Hori, C. E., Steffen Jr., V. Improved Bees Colony Algorithm Applied to Chemical Engineering System Design. *International Review of Chemical Engineering*, 2(6), 714–719, 2010.

Lobato, F. S., Souza, D. L., Gedraite. R. A Comparative Study using Bio-inspired Optimization Methods Applied to Controllers Tuning, In Frontiers in Advanced Control Systems, G. Luiz de Oliveira Serra (Editor), 2012.

Lobato, F. S., Sousa, M. N., Silva, M. A., Machado, A. R. Multi-objective Optimization and Bio-inspired Methods Applied to Machinability of Stainless Steel. *Applied Soft Computing*, 22, 261–271, 2014.

Lucic, P., Teodorovic, D. System: Modeling Combinatorial Optimization Transportation Engineering Problems by Swarm Intelligence. Preprints of the TRISTAN IV Triennial Symposium on Transportation Analysis, 441–445, 2001.

Lukasik, S., Zak, S. Firefly Algorithm for Continuous Constrained Optimization Task. ICCCI 2009, Lecture Notes in Artificial Intelligence (Eds. N. T. Ngugen, R. Kowalczyk, S. M. Chen), 5796, 97–100, 2009.

Madeiro, S. S. Modal Search for Swarm based on Density, Dissertation, Universidade de Pernambuco (in portuguese), 2010.

Nayfeh, A. H., Perturbation Methods, John Wiley & Sons, Inc., 2000.

Nissen, J. C., Popp, K., Schmalhorst, B., Optimization of a Non-linear Dynamic Vibration Absorber, *Journal of Sound and Vibration*, 99(1), 149–154, 1985.

Pai, P. F., Schulz, M. J., A Refined Nonlinear Vibration Absorber. *International Journal of Mechanical Sciences*, 42(3), 537–560, 2000.

Pareto, V. Manuale di Economia Politica, Societa Editrice Libraria, Milano, Italy, 1906. Translated into English by A.S. Schwier as Manual of Political Economy, Macmillan, New York, 1971.

Parrish, J., Viscido, S., Grunbaum, D. Self-organized Fish Schools: An Examination of Emergent Properties. *Biological Bulletin*, 202(3), 296–305, 2002.

Pfeifer, A. A., Lobato, F. S. Solution of Singular Optimal Control Problems using the Firefly Algorithm. Proceedings of VI Congreso Argentino de Ingenieria Quimica - CAIQ2010, 2010.

Pham, D. T., Kog, E., Ghanbarzadeh, A., Otri, S., Rahim, S., Zaidi, M. The Bees Algorithm - A Novel Tool for Complex Optimisation Problems. Proceedings of 2nd International Virtual Conference on Intelligent Production Machines and Systems, Oxford, Elsevier, 2006.

Prékopa, A. Stochastic programming. 1995. Dordrecht, the Netherlands: Springer.

Rice, H. J., McCraith, J. R. Practical Non-linear Vibration Absorber Design. *Journal of Sound and Vibration*, 116(3), 545–559,1987.

Ritto, T. G., Sampaio, R., Cataldo, E. Timoshenko Beam with Uncertainty on the Boundary Conditions. *Journal of the Brazilian Society of Mechanical Science and Engineering*, XXX(4), 295-303, 2008.

Rocha, A. M. A. C., Martins, T. F. M. C., Fernandes, E. M. G. P. An Augmented Lagrangian Fish Swarm based Method for Global Optimization. *Journal of Computational and Applied Mathematics*, 235, 4611–4620, 2011.

Sampaio, R., Soize, C., 2007, On Measures of Nonlinearity Effects for Uncertain Dynamical Systems Application to a Vibro-impact System. *Journal of Sound and Vibration*, 303, 659–674.

Shaw, J., Shaw, S. W., Haddow, A. G. On the Response of the Nonlinear Vibration Absorber. *International Journal of NonLinear Mechanics*, 24(4), 281–293,1989.

Shen, W., Guo, X., Wu, C., Wu, D. Forecasting Stock Indices using Radial Basis Function Neural Networks Optimized by Artificial Fish Swarm Algorithm. *Knowledge-Based Systems*, 24, 378–385, 2011.

Soize, C. Maximum Entropy Approach for Modeling Random Uncertainties in Transient Elastodynamics. *Journal of the Acoustical Society of America*, 109(5), 1979–1996, 2001.

Soize, C. A Comprehensive Overview of a Non-parametric Probabilistic Approach of Model Uncertainties for Predictive Models in Structural Dynamics. *Journal of Sound and Vibration*, 288(3), 623–652, 2005.

Souza, D. L., Lobato, F. S., Gedraite, R., Robust Multiobjective Optimization Applied to Optimal Control Problems Using Differential Evolution. *Chemical Engineering & Technology*, 38(4), 721–726, 2015.

Steffen Jr., V., Rade, D. A. Optimisation of Dynamic Vibration Absorbers over a Frequency Band. *Mechanical Systems and Signal Processing*, 14(5), 679–690, 2000.

Steffen Jr., V. and Rade, D. A. Dynamic vibration absorber. In Encyclopedia of Vibration, 9–26, Academic Press, 2001.

Taguchi, G. Quality Engineering through Design Optimization. Kraus International Publications, New York, 1984.

Tarantola, A. Inverse Problem Theory and Methods for Model Parameter Estimation, Society for Industrial and Applied Mathematics, ISBN 0-89871-572-5, 2005.

Teodorovic, D., Dell'Orco, M. Bee Colony Optimization - A Cooperative Learning Approach to Complex Transportation Problems. Proceedings of the 10th EWGT Meeting and 16th Mini-EURO Conference, 2005.

Thomsen, J. J. Vibrations and Stability, Springer-Verlag, 2nd Edition, 2003.

Tsai, H. C., Lin, Y. H. Modification of the Fish Swarm Algorithm with Particle Swarm Optimization Formulation and Communication Behavior. *Applied Soft Computing Journal*, 11, 5367–5374, 2011.

Viana, F. A. C., Kotinda, G. I., Rade, D. A., Steffen Jr., V. Tuning Dynamic Vibration Absorbers by using Ant Colony Optimization. *Computers and Structures*, 86, 1539–1549, 2008.

von Frisch, K. Bees: Their Vision, Chemical Senses and Language. Revised edn, Cornell University Press, N.Y., Ithaca, 1976.

Wang, C. R., Zhou, C. L., Ma, J. W. An Improved Artificial Fish-Swarm Algorithm and Its Application in Feed forward Neural Networks. Proc. of the Fourth Int. Conf. on Machine Learning and Cybernetics, 2890–2894, 2005.

Wong, W. O., Cheung, Y. L., Optimal Design of a Damped Dynamic Vibration Absorber for Vibration Control of Structure Excited by Ground Motion. *Engineering Structures*, 30, 282–286, 2008.

Xia, F., Zhao, X., Zhang, J., Ma, J., Kong, X. BeeCup: A Bio-inspired Energy-Efficient Clustering Protocol for Mobile Learning. *Future Generation Computer Systems*, 37, 449–460, 2014.

Yang, X. S. Nature-inspired Metaheuristic Algorithms. Luniver Press, Cambridge, 2008.

Yang, X. S. Engineering Optimizations via Nature-inspired Virtual Bee Algorithms. IWINAC 2005, Lecture Notes in Computer Science, 3562, Edited by Yang, J. M. and J.R. Alvarez, Springer-Verlag, Berlin Heidelberg, 317–323, 2005.

Yang, X. S. Firefly Algorithm for Multimodal Optimization. *Stochastic Algorithms: Foundations and Applications*, 5792, 169–178, 2009.

Yang, X. S. Firefly Algorithm, Lévy Flights and Global Optimization. In: Research and Development in Intelligent Systems XXVI (Eds M. Bramer, R. Ellis, M. Petridis), Springer London, 209–218, 2010.

Yang, C., Li, D., Cheng, L. Dynamic Vibration Absorbers for Vibration Control within a Frequency Band. *Journal of Sound and Vibration*, 330, 1582–1598, 2011.

Zhu, S. J., Zheng, Y. F., Fu, Y. M. Analysis of Non-linear Dynamics of a Two Degree-of-Freedom Vibration System with Non-linear Damping and Non-linear Spring. *Journal of Sound and Vibration*, 271(2), 15–24, 1992.

Zhu, J., Mottershead, J., Kyprianou, A. An Inverse Method to Assign Receptances by Using Classical Vibration Absorbers. *Journal of Vibration and Control*, 15(1), 53–84, 2009.

Zúniga, E. C. T., Cruz, I. L. L., García, A. R. Parameter estimation for crop growth model using evolutionary and bio-inspired algorithms. *Applied Soft Computing*, 23 (2014) 474–482.

Printed by Books on Demand GmbH, Norderstedt / Germany